MIX
Papier aus verantwortungsvollen Quellen
Paper from responsible sources
FSC® C105338

Susanne Bölke

Strategic Marketing Approaches within Airline Management

How the Passenger Market causes the Business Concepts of Full Service Network Carriers, Low Cost Carriers, Regional Carriers and Leisure Carriers to overlap

Anchor Academic Publishing

Bölke, Susanne: Strategic Marketing Approaches within Airline Management: How the
Passenger Market causes the Business Concepts of Full Service Network Carriers, Low
Cost Carriers, Regional Carriers and Leisure Carriers to overlap. Hamburg, Anchor
Academic Publishing 2014

Buch-ISBN: 978-3-95489-285-3
PDF-eBook-ISBN: 978-3-95489-785-8
Druck/Herstellung: Anchor Academic Publishing, Hamburg, 2014

Bibliografische Information der Deutschen Nationalbibliothek:
Die Deutsche Nationalbibliothek verzeichnet diese Publikation in der Deutschen
Nationalbibliografie; detaillierte bibliografische Daten sind im Internet über
http://dnb.d-nb.de abrufbar.

Bibliographical Information of the German National Library:
The German National Library lists this publication in the German National Bibliography.
Detailed bibliographic data can be found at: http://dnb.d-nb.de

All rights reserved. This publication may not be reproduced, stored in a retrieval system
or transmitted, in any form or by any means, electronic, mechanical, photocopying,
recording or otherwise, without the prior permission of the publishers.

Das Werk einschließlich aller seiner Teile ist urheberrechtlich geschützt. Jede Verwertung
außerhalb der Grenzen des Urheberrechtsgesetzes ist ohne Zustimmung des Verlages
unzulässig und strafbar. Dies gilt insbesondere für Vervielfältigungen, Übersetzungen,
Mikroverfilmungen und die Einspeicherung und Bearbeitung in elektronischen Systemen.

Die Wiedergabe von Gebrauchsnamen, Handelsnamen, Warenbezeichnungen usw. in
diesem Werk berechtigt auch ohne besondere Kennzeichnung nicht zu der Annahme,
dass solche Namen im Sinne der Warenzeichen- und Markenschutz-Gesetzgebung als frei
zu betrachten wären und daher von jedermann benutzt werden dürften.

Die Informationen in diesem Werk wurden mit Sorgfalt erarbeitet. Dennoch können
Fehler nicht vollständig ausgeschlossen werden und die Diplomica Verlag GmbH, die
Autoren oder Übersetzer übernehmen keine juristische Verantwortung oder irgendeine
Haftung für evtl. verbliebene fehlerhafte Angaben und deren Folgen.

Alle Rechte vorbehalten

© Anchor Academic Publishing, Imprint der Diplomica Verlag GmbH
Hermannstal 119k, 22119 Hamburg
http://www.diplomica-verlag.de, Hamburg 2014
Printed in Germany

Abstract

On the German passenger market, airlines approach different business concepts in order to cope with the threats and to be successful. The case of the Lufthansa Passenger Airline and its subsidiary Germanwings has been discussed many times currently. Together they have implemented a restructured concept of the Low Cost Carrier Germanwings in order to overcome their weaknesses. The purpose of this paper is to evaluate the potential of economic success of this strategy change. Therefore, the paper comprises three main areas. The first one is the theoretical part, which explains the differences between Full Service Network Carriers, Low Cost Carriers, Regional Carriers and Leisure Carriers. Secondly, the analysis takes place by applying Porter's five forces model. The results show that the German passenger airline market is of low attractiveness for airlines. Reasons for this are a strong rivalry among existing competitors, operation barriers, the existence of substitute products and services, as well as a strong bargaining power of the suppliers and a relatively strong bargaining power of the customers. Subsequently, the strengths and weaknesses of the Lufthansa Passenger Airline and Germanwings are highlighted and the new business concept is introduced. Finally, all findings are put into relation using the SWOT-analysis. In conclusion, it can be said that the new business concept combines the strengths of a Full Service Network Carrier and of a Low Cost Carrier. Nevertheless, it is doubtful whether this concept will be more successful than the old concept of Germanwings.

Contents

Abstract I

Contents III

List of Abbreviations VI

List of Figures VII

List of Tables IX

1 Introduction 2
  1.1   Research Objectives 2
  1.2   Overview of the Study 2
  1.3   Research Methodology 3

2 Strategic Airline Marketing 4
  2.1   The Classical Marketing and the Marketing Management Process 4
  2.2   The Marketing Mix Applied by Airlines 6
      2.2.1 *Product Policy* 6
      2.2.2 *Price Policy* 10
      2.2.3 *Place Policy* 12
      2.2.4 *Promotion Policy* 14

3 Applied Business Concepts of Airlines 16
  3.1   Full Service Network Carriers 17
      3.1.1 *Marketing Mix* 18
      3.1.2 *Advantages and Disadvantages* 19
  3.2   Low Cost Carriers 20
      3.2.1 *Marketing Mix* 20
      3.2.2 *Advantages and Disadvantages* 21
  3.3   Regional Carriers 21
      3.3.1 *Marketing Mix* 22
      3.3.2 *Advantages and Disadvantages* 22
  3.4   Leisure Carriers 23
      3.4.1 *General Characteristics* 23
      3.4.2 *Marketing Mix* 23
  3.5   Overlapping Business Concepts 24

4 Analysis of the German Passenger Airline Market – Porter's Five Forces 26

4.1 Rivalry among Existing Competitors 27
- 4.1.1 Competition among Low Cost Carriers 29
- 4.1.2 Full Service Network Carriers and Strategic Alliances 31
- 4.1.3 Differentiation of Products and Services 32
- 4.1.4 The German Passenger Airline Market in Europe 32

4.2 Threat of New Entry 33
- 4.2.1 Entry Barriers through Slot Allocation Regulations 33
- 4.2.2 The Liberalization of European Air Traffic 35
- 4.2.3 Subsidies at Regional Airports 36
- 4.2.4 The Elimination of Further Subsidies via the Aviation Tax Act 37
- 4.2.5 Competitive Disadvantages Through Emission Trading 38
- 4.2.6 Relatively High Capital and Resource Requirements 39

4.3 Threat of Substitute Products and Services 40
- 4.3.1 Intermodal Competition by Road and Rail 40
- 4.3.2 Substitution Through New Media 42

4.4 Bargaining Power of Suppliers 43
- 4.4.1 The Power of Aircraft Manufacturers 43
- 4.4.2 The Bargaining Power of Airports 44
- 4.4.3 Global Distribution Systems 45

4.5 Bargaining Power of Buyers 45
- 4.5.1 The Number of Buyers 46
- 4.5.2 Switching Costs 46

4.6 Analysis of the Results of Porter's Five Forces 47

5 The Demand within the German Passenger Airline Market 50

5.1 Business and Private Traveler in the Decision-Making Process 50
5.2 Development of the Passenger Demand in Germany 55
- 5.2.1 Development from 2001 to 2013 and Forecast 55
- 5.2.2 The Development of Business and Private Travel 58

6 The Case of Lufthansa and Germanwings 60

6.1 The Lufthansa Group 60
6.2 The Lufthansa Passenger Airline Group and its Business Performance 61
6.3 Business Performance of the Lufthansa Passenger Airline 64
6.4 The Subsidiary Germanwings and its Business Performance 67

7	The SWOT-Analysis applied for Lufthansa and Germanwings	70
7.1	The New Germanwings – a Strategy Change	73
7.2	Evaluation of the New Strategy in Accordance to the SWOT-Analysis	76
8	**Conclusion and Recommendations**	**80**

Appendix **83**

List of References **91**

Books, Book Sections and Videos 91

Magazine, Newspaper and Journal Article 92

Reports, Documents, Presentations, Emails 92

Websites 96

List of Abbreviations

ASK	Available Seat-Kilometers
FSNC	Full Service Network Carrier
GDP	Gross Domestic Product
GDS	Global Distribution System
GW	Germanwings
LC	Leisure Carrier
LCC	Low Cost Carrier
LH	Lufthansa
NGW	New Germanwings
RC	Regional Carrier
RPK	Revenue Passenger-Kilometers
SWOT	Strengths, Weaknesses, Opportunities, Threats

List of Figures

Figure 1:	Overlapping Airline Business Concepts	24
Figure 2:	Michael Porter's Five Forces Model	26
Figure 3:	Market Share of Business Concept in Germany in 2011	29
Figure 4:	Route Network offered by LCC in Germany (January of each year)	30
Figure 5:	LCC Market Share in Germany in January 2013	31
Figure 6:	FSNC and Strategic Alliances Market Share in Germany in 2011	32
Figure 7:	Geographical Segmentation of the European Airline Industry Value	33
Figure 8:	Attractiveness of the German Passenger Airline Market	47
Figure 9:	Criteria for Transport Mode Selection	53
Figure 10:	Decision-Making Criteria for Booking a Business Trip	54
Figure 11:	Decision-Making Criteria according to Preferences	55
Figure 12:	Annual Percentage Change of Passenger Volume (Arrivals and Departure) of Schedule and Charter Flights at Germany's 22 most Congested Airports	56
Figure 13:	Passenger Volume at German Airports According to Destination and Origin Respectively	57
Figure 14:	Portions of Business and Private Travelers at the Cologne-Bonn Airport	59
Figure 15:	Total Numbers of Business and Private Travelers at the Cologne-Bonn Airport	59
Figure 16:	Lufthansa Passenger Airline Group - ASK and RPK from 2007 to 2012	63
Figure 17:	Lufthansa Passenger Airline Group - Operating Result from 2007 to 2012	64
Figure 18:	Lufthansa Passenger Airline - Number of Passengers	65
Figure 19:	Lufthansa Passenger Airline - ASK and RPK in Millions	66
Figure 20:	Lufthansa Passenger Airline - Operating Result	66

Figure 21:	Germanwings - Number of Passengers from 2005 to 2011 in July	67
Figure 22:	Germanwings - Load Factor from 2005 to 2011 in July	68
Figure 23:	Germanwings - Operating Result	69
Figure 24:	Development of the GDP in Germany and Forecast	83
Figure 25:	Market Share of Existing Business Concepts	84
Figure 26:	FSNC Supply According to Alliance Membership	85
Figure 27:	The Lufthansa Group Business Segments - Share of the Operating Result and the Revenue in 2012	86
Figure 28:	Operating Result of the Lufthansa Group's Business Segments	87
Figure 29:	Lufthansa Passenger Airline Group - Number of Passengers	88
Figure 30:	Lufthansa Passenger Airline Group - Load Factor from 2007 to 2012	88
Figure 31:	Lufthansa Passenger Airline Group - Operating Expenses in m€	89
Figure 32:	Lufthansa Passenger Airline Group - Regional Market Share	90

List of Tables

Table 1:	Classification of Airlines Operating in Germany	28
Table 2:	Profiles of Private Travelers	52
Table 3:	Business Segments of The Lufthansa Group	60
Table 4:	SWOT-Portfolio Lufthansa Passenger Airline and Germanwings	71
Table 5:	The Business Concept of the NGW	75
Table 6:	Geographical Segmentation of the European Passenger Airline Market in 2011	83
Table 7:	Market Share of Existing Business Concepts	84
Table 8:	FSNC Supply According to Alliance Membership	85
Table 9:	Lufthansa Passenger Airline Group - Business Performance	87
Table 10:	Lufthansa Passenger Airline Group - Operating Expenses in m€	89
Table 11:	Germanwings - Business Performance	90

1 Introduction

According to SHAW the success of an airline depends a lot on the applied marketing mix, which is the result of strategic marketing and the business concept of an airline.[1]

Through the commencements of new acts, the liberalization of the passenger market as well as a change in demand, the market has always been in motion. New airlines have emerged and existing ones have disappeared due to different opportunities and threats. Consequently, the German passenger airline market has changed during the last decade from a Full Service Network Carrier (FSNC) shaped to a more and more Low Cost Carrier (LCC) dominated one.

The most current case for changing the business concept according to the recent market situation on the German passenger airline market, is the case of the Lufthansa Passenger Airline and its subsidiary Germanwings.

1.1 Research Objectives

Airlines apply different marketing strategies and thus different marketing mixes in order to tackle the everyday challenges of the German passenger airline market. The objectives of this paper are to give an overview of the recent situation of this market, and to prove the potential for success for the new business concept applied by the German airlines Lufthansa Passenger Airline and Germanwings in the sense of overcoming their weaknesses and to take roots.

1.2 Overview of the Study

In order to reach the purpose of this paper, the inductive research method is applied, which observes the general market and draws a conclusion at the end. Thus, the paper is divided into three major parts, which are the theoretical one, the analyses and the evaluation part.

---

[1] Shaw (2011), p. 6.

Chapter two jumps directly into the theoretical section. It comprises information about strategic marketing in general and the marketing mix in particular. Furthermore, it applies the marketing mix to the passenger airline market. The third part describes the existing business concepts of airlines as a result of applied strategic marketing. It evaluates them according to different features and their potential success.

After the completion of the theoretical part, different analyses are applied in order to state the status quo of the German passenger airline market. In chapter four Porter's five forces model is used to state the recent situation of the market itself. This tool gives the opportunity to evaluate the competition amongst the existing airlines, but also focuses on indirect competitors, which may also minimize the profitability of an airline. However, this tool only assesses the macro-environment of an airline and does not include information about the demand and the micro-environment. Consequently, further analyses are approached in the subsequent chapters. The recent demand and a forecast for the next years are discussed in chapter five and the performances of the Lufthansa Passenger Airline and Germanwings during the previous years are stated in chapter six.

Chapter seven combines the results of the aforementioned analyses in order to evaluate the findings and the just implemented business concept of Lufthansa and Germanwings. Finally, a conclusion is drawn and some recommendations are given.

1.3   Research Methodology

Secondary research was done on quantitative as well as on qualitative data according to each part. Thus, the theoretical part is mainly based on literature reviews. For the analysis section, statistics were used additionally. These comprise statistics by German and European organizations (Arbeitsgemeinschaft Deutscher Verkehrsflughäfen, Verband Deutsches Reisemanagement e.V. Deutsches Zentrum für Luft- und Raumfahrt e.V., etc.), suppliers (e.g. AIRBUS), airlines (Lufthansa, Germanwings, etc.), and the government.

2 Strategic Airline Marketing

2.1 The Classical Marketing and the Marketing Management Process

The meaning of the term 'marketing' has developed during the last 60 years. Originally, it used to describe the promotion and distribution of products only. Later the satisfaction of consumer needs in order to reach the company's economic objectives became important. Next, a more complex process, including planning, coordination and control, was taken into account.[2] Thus, a variety of definitions of 'marketing' with different core statements exist today. However, today's markets are very complex and each component mentioned might lead to success or to failure to the same degree. In addition, many components overlap and can influence each other. Thus, another definition is needed. The AMERICAN MARKETING ASSOCIATION defines marketing as "the activity, set of institutions, and processes for creating, communicating, delivering, and exchanging offerings that have value for customers, clients, partners, and society at large."[3] This definition appears to be the most current and most appropriate, since it covers the main facets of marketing mentioned above, and can be applied to products as well as services. Additionally, it does not only take into account customers and the company, but all existing stakeholders.

On the basis of BRUHN and MEFFERT, the marketing management process can be divided into four different main phases:

- the analysis of the current situation
- the planning phase
- the implementation phase
- the controlling phase

---

[2] Meffert (2012), p. 10-12.
[3] American Marketing Association (2007).

During the first stage, the macro environment is analyzed. Thus, all external factors, which have or might have an impact on the company's performance, are identified. Additionally, the branch which the company is operating in is inspected. Therefore, research is done on the competition, the target groups, as well as on the supply. Last, but not least, the company's current situation is analyzed and a future trend is given.

According to the research results of the first phase, marketing objectives can be defined. These objectives can be economic, such as the achievement of a certain turnover or a certain profit margin, or pre-economic, such as the change of the demand behavior.[4] Additionally, MEFFERT mentions that marketing objectives might have a social or an ecological background as well.[5] However, these objectives can also be considered as pre-economic objectives since by achieving them, companies again try to improve some economic results, such as the profit. Hence, the marketing objectives have to conform to the company's overall objectives in order to be successful.

According to the set objectives, a marketing strategy is formulated in the next step. Among others, this includes the definition of the target segments and the target groups, which are supposed to be focused on in the future, as well as a framework for the marketing plan. On the basis of the marketing strategy the operative marketing mix can then be defined. This includes decisions on the product, the price, the promotion and the place/distribution (4 Ps).[6] Different authors also extend this classical marketing mix of the four Ps to seven Ps in the service sector. Thus, additional decisions are made on personnel, processes and physical facilities.

The third phase of the marketing management process is the implementation phase. During this phase, decisions on how to implement the marketing mix have to be made. These include, for example, decisions on the distribution of responsibilities to the relevant departments and/or to certain people, decisions on the time frame, as well as decisions on the budget. As soon as all these decisions are made, the marketing mix can be realized. The last stage of the marketing management process, phase four, is the controlling phase. At this

---

[4] Bruhn (2012), p. 37 ff.; Meffert (2012), p. 20.
[5] Meffert (2012), p. 21.
[6] Bruhn (2012), p. 47-49; Meffert (2012), p. 21.

point, the achieved results match the before set objectives. Although controlling is stated as the last stage of the process, it needs to be done along all phases in order to be able to react to sudden changes at any time.[7]

Each stage of the marketing management process can be sub-divided further; however, this paper is intended to analyze whether the current applied business concepts of airlines are strategically useful according to the market situation. As a consequence, only the marketing mix, which includes the operative tools product, price, place and promotion, will be broken down further.

2.2 The Marketing Mix Applied by Airlines

According to CONRADY, FICHERT and STERZENBACH, the marketing mix can be seen as the 'heart' of the marketing management process.[8] Additionally, SHAW describes the classical marketing mix as a 'powerful model' to be a successful airline.[9] As a consequence, the four Ps (product, price, place, and promotion), or rather the airlines' business concepts, play a major role for airlines and shall, thus, be described further in the following part of this chapter.

2.2.1 Product Policy

The product policy describes the way of analyzing the current products and the design of new ones. This includes all decisions that have to be made in order to establish a product or a service which can be offered on the market later on.[10]

The classic product is a combination of different characteristics with the objective to satisfy certain customer needs. It can be tangible or non-tangible in the sense of services. As a result, the word 'product' in this paper always refers to services as well. In the classic marketing the product usually consists of the core product, which is supposed to fulfill basic needs only, and the augmented product, which is not essential but adds value to the product in order to satisfy

---

[7] Bruhn (2012), p. 47-49; Meffert (2012), p. 21.
[8] Conrady, Fichert, and Sterzenbach (2013), p. 418.
[9] Shaw (2011), p. 6.
[10] Meffert (2012), p. 385.

additional needs.[11] This description can be applied to tangible products, such as cars, easily. Here, the core product is the car, which fulfills the basic function of driving from destination A to destination B, and the augmented product could be the interior of the car, which is not essentially needed, but adds value to the product. This approach can be applied to services, such as hotel accommodations, as well. In this case, the core product is the night spent in the hotel itself, which meets the customer's need of lodging. The augmented product could be a mini bar in the room, which again adds value to the hotel night by offering something that is not essential but nice to have. The combination of the core product and the augmented products allows the creation of unique products and, thus, the distinction from other products.

However, this product approach cannot be applied to any service. Airlines, for example, are not able to distinguish as strictly between the core product and the augmented product. Here the core product comprises the basic flight from destination A to destination B. Thus, the basic need of the customer is met. But there are many services coming along with the booking process already, such as the mode of payment, for example. These services are essential but not a part of the core product. However, they already allow for a differentiation from other products but yet are not augmented products. The customer has to purchase these services in order to use the core product. Augmented products of airlines are, for example, snacks, which are provided during the flight, or movies which can be watched on board of the aircraft. These services are not essential for the realization of the flight but add value to it.[12]

The next section of this paper will give insight into the variety of components the airline product may comprise in order to achieve a sustainable competitive advantage. The main elements are the route network, the type of aircraft, the flight frequencies, and timings.[13] Alongside a variety of other features are available. In relation to the point of realization, these can be divided into three different groups – pre-flight, in-flight, and post-flight features, which finally form the service chain. Pre-flight features are those that are used starting with the

---

[11] Meffert (2012), p. 387.
[12] Conrady, Fichert, and Sterzenbach (2013), p. 419.
[13] Shaw (2011), p. 5.

booking process to the point of boarding. They include services such as reservation, booking, payment, transfer to the airport, entrance to the terminal, and services that are offered at the airport (check-in, lounges, etc.). In-flight features are those realized during the flight, such as the catering, restrooms, and entertainment. After de-boarding, post-flight services can be used. These are, for example, the transfer to the next airplane, the transfer within the airport, a security check or a transfer to the final destination. All mentioned services may be components of the product or might be purchased separately depending on the product design of each airline. Additionally it needs to be said that the stated features only give an insight into the variety of services available. The list of existing services is much longer and is extended constantly.[14]

The product planning can be divided into five categories, on which decisions have to be made. These are product innovations, product variations, product differentiations, product eliminations, and product diversifications.
Innovations describe the creation of new products and services, which have not previously been on the market yet. Product innovations, if successful, are known to be the most profitable ones. However, there is always a chance of failing. Additionally, the creation of new products is costly in terms of resources needed.[15] According to the mentioned definition, product innovations are fairly rare in the passenger airline industry, though CONRADY, FICHERT and STERZENBACH argue that the product of the LCC, which is described in chapter 3.2, is an innovation.[16] However, this is an entirely new concept comprising more characteristics than those of a single product. It takes into account all of the four Ps at once. Thus, the concept itself cannot be seen as a product innovation, but the product of this concept may. It is an innovation since it only combines the core product, which is the flight, and services, which are essential for flying, such as the booking process, for example.

---

[14] Conrady, Fichert, und Sterzenbach (2013), p. 422-425; Shaw (2011), p. 190-197.
[15] Meffert (2012), p. 396-399.
[16] Conrady, Fichert, und Sterzenbach (2013), p. 420.

Additional innovations, which were introduced to the German market lately, are, for example, the provisioning of internet access on board of the aircrafts.[17]

Product variations describe the change of characteristics of existing products in order to react to market changes or to eliminate product errors, whereas the product in general stays the same. Consequently, the product remains up-to-date and satisfies the current demand. By modifying the product, the number of products stays the same.[18] Within the passenger airline market products are modified regularly by adjusting flight schedules, varying catering components, etc.[19]

Another way of changing the product portfolio is the product differentiation. Here, an additional product extends the existing portfolio in order to saturate different market segments than the ones that are already served. Product differentiations are, like product variations, used to react to market changes and to make profit of arising potential, such as new target groups for example.[20] Within the passenger airline market booking classes are a well-known example for a product differentiation, where different target groups are addressed, such as business travelers via the business class and leisure travelers via the economy class. The offered classes may, for example, vary in the size of the seats, the catering or the media provided during the flight according to what the targeted customer requires.[21]

Furthermore, products can be eliminated from the existing product portfolio. Reasons for this might be, for example, the insufficient congruity with the current demand or high costs.[22] The product elimination can thus be seen as the opposite of the product diversification. Consequently, the example which was used for describing product differentiation can also be used for describing

---

[17] Opfermann (2012).
[18] Meffert (2012), p. 446 f.
[19] Conrady, Fichert, and Sterzenbach (2013), p. 420.
[20] Meffert (2012), p. 446.
[21] Conrady, Fichert, and Sterzenbach (2013), p. 421.
[22] Meffert (2012), p. 454.

product eliminations within passenger airline market.[23] Here different booking classes can be eliminated in order to stop addressing a certain target group.

Contrary to MEFFERT's four components, CONRADY, FICHERT and STERZENBACH mention a fifth component, which a decision can be made on - the product diversification. It also describes the expansion of the current product portfolio by products which are either related to the existing ones (horizontal diversification), which are a part of the supply chain (vertical diversification), or which are not related to the existing products (lateral diversification). All three kinds of product diversification can be applied by airlines.

Air cargo, for example, is a case of horizontal diversification. Here the airline offers an additional product, which is similar to the existing ones, but can use conjunct resources, such as the aircrafts. Purchasing a catering company would be an example of vertical diversification. When a company, which usually operates in a totally different field of business, starts to operate in the passenger airline market, it is considered a case of lateral diversification.[24]

2.2.2 Price Policy

The price policy covers all decisions that have to be made on the fare that is charged for the final product. This includes, for example, decisions on discounts as well as on payment and delivery conditions in accordance with the overall marketing strategy of the company.

In most cases price policy aims to maximize the profit. However, there is a range of other objectives that companies try to achieve nowadays. These comprise objectives regarding the company itself, such as an increased cost recovery, full employment or a stable position in the market, and objectives regarding the position in the market in general, such as the winning of new market segments and new customers, or the creation of a certain image.[25]

---

[23] Conrady, Fichert, and Sterzenbach (2013), p. 420.
[24] Conrady, Fichert, and Sterzenbach (2013), p. 420-425.
[25] Meffert (2012), p. 466-470.

The price policy offers a wide range of strategies and instruments to find the right price for the offered products. However, not all of them can be used at any time to the same degree for any product or service due to different characteristics. This also applies for airlines. Their products consist of typical service characteristics. They are, for example, not storable and the production and the consumption take place at the same time, which means that the core product expires after the production, which in this case is the flight. The price building process of airlines therefore is very complex and depends on many factors, such as the point-of-sale, the target group, the season, etc. However, to keep it simple, it can be said that in general the price policy process can be divided into two phases. The pricing process, which indicates the fare levels for each product, and the revenue management, which states how many seats are to be sold at every level. Since this paper looks at the status quo of airlines business concepts and its effect on the customers only, it will not consider dynamic pricing strategies and the revenue management process in detail. Furthermore, it will not take into account, whether prices are based on the costs, the competition, or the value.[26]

The pricing process of airlines can be divided into two main stages. First of all a price level has to be set. It has to be decided whether prices are supposed to be on a low, a medium or a high level. The low price strategy is also known as promotion price strategy. It is usually used to sell a simple product with a few added services only. This strategy usually does not need much promotion since the low prices are promoting the product themselves. The high price strategy, in comparison, also known as premium price strategy, requires more promotion. This strategy is mainly used for selling more complex products, which include the core product as well as a range of augmented products. All price levels in between are known as medium price strategies.[27] Some authors argue that airlines mainly apply the two major strategies; the premium price strategy or the promotion price strategy.[28] However, today such a strict separation is nearly

---

[26] Conrady, Fichert, and Sterzenbach (2013), p. 362; Belobaba (2009), p. 73.
[27] Meffert (2012), p. 493.
[28] Conrady, Fichert, und Sterzenbach (2013), p. 354.

impossible since the price levels vary just as the products of the airlines do. There are prices available at a low, a medium and a high price level as well.

Additionally, the "No Frills Concept", also known as the "A La Carte Pricing", has become more important recently. It allows the airline to sell the core product (the flight from A to B) separately and to achieve additional revenues by selling further services. These revenues are called ancillary revenues (see chapter 2.2.1).

Once a decision on the price level is made, a price differentiation takes place. This means that the same product is offered for different prices at the same time. Through price differentiation, the pricing process can get very complex depending on the overall business concept of the airline. Price differentiation is done in order to skim the highest possible consumer's surplus. However, the willingness to pay varies from target group to target group depending on different factors. Thus, a differentiation within the passenger airline industry is usually done related to the flight schedule, the location where the flight is purchased, the customer status (student, business traveler, leisure traveler, etc.), and the quantity of flights purchased.[29]

2.2.3 Place Policy

The place policy covers all decisions that have to be made regarding the distribution of the products from the producer to the final customer or the intermediary. Thus, the word 'place' can also be replaced by the word 'distribution'.[30]

In the context of the place policy an entire distribution system is designed, whereas the final design depends on a number of different factors against the background of the objectives and the concept of the business. These are, for example, the availability to the customers, the height of the costs, the

---

[29] Conrady, Fichert, und Sterzenbach (2013), p. 354 f; Meffert (2012), p. 499 f.
[30] Meffert (2012), p. 543.

controllability as well as the adaptability of the distribution channel and the achieved yield by each product sold.[31]

The design process can be divided into two steps. First of all, decisions have to be made on the vertical structure. It has to be decided whether products are supposed to be sold directly to the final customer, or whether one or several intermediaries are supposed to be in between (indirect distribution). In the case of indirect distribution, further decisions have to be made on the number of intermediaries, on the kind of intermediaries, and on contractually conditions if necessary.

Secondly, decisions need to be made on the horizontal structure of the distribution channel, which includes decisions on the width of each step (intensive, selective and exclusive market concentration) and the depth of each step of the distribution channel (type of business).[32]

Airlines use different ways to sell their products to their customers. Direct distribution mainly happens via websites, call centers, ticket offices at airports and cities as well as via selling points for employees. On the other hand, indirect distribution mainly takes place via travel agencies (online and offline), tour operators, and consolidators.[33]

Global Distribution Systems play a major role for airlines. These are computer based reservation systems, which combine products and services of different service providers, such as airlines, hotels, and car rentals. They are operated independently and on a worldwide basis. These systems enable the user to compare the products offered by different suppliers and to purchase them easily. Yet, they make the distribution of tickets easy, they also make them more expensive, since the airlines have to pay fees in order to use them.[34]

---

[31] Conrady, Fichert, und Sterzenbach (2013), p. 439 f.
[32] Meffert (2012), p. 550.
[33] Meffert (2012), p. 441-445.
[34] Schulz (2010), p. 264-265.

2.2.4 Promotion Policy

The promotion policy covers all decisions that have to be made in order to communicate information about the company and its products, to increase the awareness and to achieve the company's overall objectives in return.

Within the promotion policy a separate strategy has to be formulated according to the overall communication objectives. The strategy includes, for example, decisions on the communication budget, such as the amount and its distribution, as well as the communication instruments on which the budget shall be spent.
Since this paper is supposed to evaluate the final result of the promotion policy of an airline, which is a part of its business concept, it will have a further look at the most common communication instruments, which matter to airlines most, only.[35]

The Corporate Identity plays a significant role in the promotion policy of an airline. The term refers to the identity of a company, which is supposed to express the company's portrait externally and internally based on a pre-defined mission and a pre-defined image. It comprises the following items:

- the Corporate Design, which describes the visual appearance,
- the Corporate Communication, which describes the usage of different communication instruments,
- the Corporate Behavior, which describes the working behavior of all employees, and
- the Corporate Culture, which comprises norms and values defined by the company.

The communication in general can be divided into indirect and direct communication. Indirect communication happens via tools, such as classical media advertising (e.g. television and radio commercials, advertising on print media), sponsoring (e.g. of people, soccer clubs, organizations), or online

---

[35] Meffert (2012), p. 606 f.

advertising. Whereas indirect communication is available to a high number of people and addresses the audience by chance, direct communication reaches a smaller amount of people, but can ensure to reach them by addressing them directly. Direct communication includes instruments, such as newsletter, dot mailer, and telephone marketing.

Furthermore, frequent flyer programs play a significant role within the passenger airline market. These programs are instruments of the customer relationship marketing, which reward the traveler for flying with the airline. These programs are mainly used by FSNC but have enjoyed more popularity among other carriers as well, lately.[36]

The list of communication instruments applied within the passenger airline market could be extended indefinitely. However, in order to draw a picture of the promotion policy within the business concept of an airline, the named items shall be sufficient.

In order to create a successful marketing mix, all components are designed in accordance with each other. They are all set to harmonize and to fulfill common objectives. Now that the processes of creating the components were described in short, this paper will have a look at the status quo of existing airline marketing mixes in the sense of business concepts.

---

[36] Conrady, Fichert, and Sterzenbach (2013), p. 453-463.

3 Applied Business Concepts of Airlines

Airlines apply business concepts according to their marketing strategy. These concepts differentiate in their marketing mix. They may have different product, different price, different distribution and/or different promotion strategies. The following pages will demonstrate the existing business concepts and their characteristics. However, it needs to be said that there is a wide range of concepts defined according to different criteria. According to these criteria CONRADY, FICHERT and STERZENBACH distinguish between three groups, which vary in their capacities as well as their flight plans. The first group consists of Full Service Network Carrier, Regional Carrier (RC), Leisure Carrier (LC) and Low Cost Carrier. These carriers operate on a fixed schedule and the seats are mainly sold separately.[37] RUPERTI as well as DOGANIS, in comparison, separate this group into three components only. They do not classify RC as a separate business concept. However, since these carriers distinguish from the other concepts in some crucial criteria, they shall be stated as their own business concept.[38]

The second group of carrier includes the concept of Business Aviation and Executive Charter. These airlines operate upon demand and all seats of one aircraft are sold to one client. The third group only contains one business concept, which is called Air Taxi or General Aviation. These carriers operate on demand and the seats of an aircraft are sold separately.[39]

The four carriers of the first group are known as the classical business concepts.[40] They dominate the German market[41] and shall, thus, be taken into further account only. The second and third groups have become more important during the last years[42] and are named for the sake of completeness, but they do not play a further role for this paper.

---

[37] Conrady, Fichert, und Sterzenbach (2013), p. 224-226; Pompl (2007), p. 104.
[38] Ruperti (2012), p. 60 f; Doganis (2010), p. 131 f.
[39] Conrady, Fichert, and Sterzenbach (2013), p. 224-226; Pompl (2007), p. 104.
[40] Ruperti (2012), p. 61.
[41] Deutsches Zentrum für Luft- und Raumfahrt e.V. (2012).
[42] Conrady, Fichert, and Sterzenbach (2013), p. 225.

Until the nineties, within the first group of carriers only three major business concepts were mentioned – Full Service Network Carriers, Regional Carriers and Leisure Carriers. Each concept was homogeneous in itself and, thus, easy to differentiate.[43] Later on another business concept occurred, the concept of Low Cost Carriers. It was still easy to separate from the others. However, a strict separation of today's existing business concepts is almost impossible, since the characteristics overlap in many cases. Some authors even say that the business concept of the LCC can be differentiated again into a low fare concept and a no frills concept. Whereas the low fare concept focuses on the price, but still adds augmented services to the basic product, and the no frills concept focuses on the basic product only.[44] Yet, this paper is supposed to focus on the classical concepts only in order to create a clear picture of the basic concepts.

3.1 Full Service Network Carriers

Full Service Network Carriers can be divided into Mega Carrier or Major Carrier, Continental Carrier, and Flag Carrier according to their market position and to their political status. Mega or Major Carriers are the market leaders seen from a commercial point of view. They are operated privately and serve intercontinentally. Some organizations even state a certain turnover, which has to be achieved by an airline in order to reach the status of a Mega Carrier. They are the leader of global alliances[45]. Flag Carriers, in comparison, are partially held by the government, which in return can implement its politics and achieve different goals, such as the provision of jobs or different routes. Through subsidies, Flag Carriers are usually able to hold their leading position, even though they are most likely not known for achieving their market leadership through their operations. Continental Carriers are usually smaller and operate on selected routes only. They play a secondary role for global alliances.[46]

---

[43] Groß and Schröder (2005), p. 11.
[44] Gross (2007), p. 13.
[45] A strategic global alliance is a cooperation of different airlines. Its goal is to achieve better results in pre-defined business fields (e.g. frequent flyer programs, code sharing). [Conrady, Fichert, and Sterzenbach (2013), p. 278 f.].
[46] Conrady, Fichert, and Sterzenbach (2013), p. 229; Pompl (2007), p. 101 f; Ruperti (2012), p. 104 f.

The concept of FSNC is known as the oldest airline business concept. Most of the biggest airlines worldwide, such as Emirates, Lufthansa, or United Airlines, are operating as FSNC. They usually serve line hauls in hub-and-spoke-systems[47] to domestic, continental, and intercontinental central airports, which are located in big cities and offer intermodal connections as well. Besides, some FSNC approach to regional airports, too. The aircrafts are usually built by the two market leaders Airbus and Boeing and offer a capacity ranging from 130 to 800 seats. The fleet of an FSNC is known to be heterogeneous due to different distances, which need to be flown, and shifts in demand, which need to be tackled. Business and private travelers are their target groups, whereas business travelers build the core target group.[48]

3.1.1 Marketing Mix

In addition to the basic product, which is the transportation from destination A to destination B, FSNC offer different products and services along the value chain attracting their target groups. These services include, for example, catering and the transportation to the airport. For this reason they are not only named Network Carriers but also Full Service Network Carriers.
On continental routes they usually offer two different booking classes, which are the business and the economy class, whereas on intercontinental flights a third (first class) and sometimes even a fourth class is added.[49]

According to the products and services added and the different booking classes a variety of different prices is available. The price per flight per person depends on factors, such as the point of time of booking and the point of time of travel, the target group and the chosen distribution channel. Usually the prices range on a higher level and comprise most of the components included in the product. Thus, no extra fees occur. However, it has also become common that FSNC try

---

[47] Two different types of route networks exist. The hub-and-spoke-system, which has one central airport, the so called 'hub', through which all remaining airports of the system, the so called 'spokes', are connected. And the point-to-point-system, which also includes several airports, which are all connected with each other through direct flights. In comparison to the hub-and-spoke-system, the number of routes within the point-to-point-system is higher [Conrady, Fichert, and Sterzenbach (2013), p. 200 f.].
[48] Conrady, Fichert, and Sterzenbach (2013), p. 226-230.
[49] Pompl (2007), p. 105; Conrady, Fichert, and Sterzenbach (2013), p. 228.

to achieve additional revenues through offering extra services, which are not part of their product.

FSNC use many channels to distribute their products. These are mainly Global Distribution Systems, which are used by travel agencies and travel management companies. Furthermore, they sell their products via the internet and call centers. Their objective is to achieve a high availability of their products and to attract all people belonging to their target group.

Communication takes place online as well as offline. FSNC also use a variety of methods to promote their products. There is no particular approach to be named. Additionally, they offer frequent flyer programs.[50]

3.1.2 Advantages and Disadvantages

FSNC usually benefit from strong brands, which they have developed over years through their economic or their political status. Besides, they usually dominate in their hubs. Consequently, FSNC have strong negotiating power, and high market awareness. Additionally, they are able to bind their customers by frequent flyer programs. Through their heterogeneous fleet they are able to adjust their supply according to the current demand in the long run. However, since FSNC usually have a fixed schedule, which they have to adhere to by law, they cannot adjust to a change in demand immediately as LC can do. Additionally, their heterogeneous fleets lead to high maintenance costs and to complex processes.

FSNC usually target both business and private travelers with a variety of different products. This mixture again leads to complex processes, which come along with high costs affecting the fares.[51] Thus, FSNC have to cope with high cost and the emergence of new business concepts, which offer the same core product for less money, such as LCC (sub-chapter 3.2).

---

[50] Pompl (2007), p. 105; Conrady, Fichert, and Sterzenbach (2013), p. 229.
[51] Conrady, Fichert, and Sterzenbach (2013), p. 227-230.

3.2 Low Cost Carriers

The business concept of LCC is the newest one. It was first approached in the eighties by an American airline and introduced to the European, including the German, market during the nineties. LCC have gained in importance ever since.[52]

The idea of LCC is to offer a basic flight from destination A to destination B in a point-to-point-system excluding additional services, such as catering and luggage, at low fares. They usually operate continentally but might offer domestic flights as well. Their fleet usually consists of one aircraft type only, offering a capacity of 150 to 250 seats each. They mainly approach regional airports, which are located close to metropolitan regions. LCC target private travelers, but due to an increasing price sensitivity of business travelers, this target group has become more important for them during the last years as well.[53]

3.2.1 Marketing Mix

The product of LCC includes, as already mentioned, the basic flight from destination A to destination B only. Services, such as luggage, catering and child entertainment are not included, but can be purchased separately. Thus, the clients themselves can decide what they want to be included and do not need to pay for services, which they do not use. The aircrafts usually have one transportation class only and seats are arranged narrow to each other.

LCC mostly differentiate from other carriers through their low prices. They offer low fares, which cover the basic flight only. Additional services, such as costs for the chosen mode of payment, luggage, catering or the reservation of a seat are not covered by this price, but can be purchased additionally. Thus, LCC are able to offer a low price, but can still achieve additional revenues through selling additional services. These revenues are called ancillary revenues.

---

[52] Doganis (2010), p. 131.
[53] Conrady, Fichert, and Sterzenbach (2013), p. 236; Pompl (2007), p. 106 f.

Furthermore, flights are only available for one price at a time but it changes over time. The closer the time of departure, the higher the price gets.

In order to reduce costs, products and services are usually sold directly via their own website and/or via their own call centers. Additionally, promotion activities are fairly simple but usually state the main message of having low fares available.[54]

3.2.2 Advantages and Disadvantages

LCC achieve their business objectives mainly through comparative cost advantages against their competitors. These cost advantages are realized through a variety of different factors. They operate with homogeneous fleets and narrow seating within the aircrafts. The products are kept very simple and there is only one price available at the time. Furthermore, LCC approach to smaller airports, where the airport fees are low and the ground handling is fast due to little services and little complexity (e.g. a small number of airlines). More costs are saved due to direct distribution. Thus, retailers do not have to be paid and complex distribution channels do not have to be maintained. The same applies to promotion activities, which are kept simple (e.g. no frequent flyer program). Nevertheless, the business concept of the LCC also has some disadvantages. Whereas a homogeneous fleet may save costs on the one side, it is very hard to react to shifts in demand on the other side due to a fixed seating capacity as well as a fixed mileage that can be flown by the aircrafts only.[55]

3.3 Regional Carriers

Regional Carriers usually operate in a point-to-point-system connecting destinations with less traffic. Additionally, they serve as feeder for FSNC (e.g. Lufthansa CityLine serves Lufthansa) by functioning between decentralized airports and their hubs. RC have a fleet of fewer and smaller aircrafts than FSNC since they only operate domestically and continentally. Each aircraft

---

[54] Conrady, Fichert, and Sterzenbach (2013), p. 239 f.; Pompl (2007), p. 110 f.
[55] Conrady, Fichert, and Sterzenbach (2013), p. 241 f.

offers a capacity ranging from 19 to 120 seats. Thus, they can also approach regional airports. Mainly business travelers use this kind of carrier.[56]

3.3.1 Marketing Mix

Products and services offered by RC depend on the size of the aircraft, which again depends on different factors, such as the demand on a certain route or the length of it. On smaller aircrafts only one booking class is available, whereas on bigger aircrafts up to two are provided. These products and services are priced at a higher level and mainly distributed via call centers or the carrier's own website. Promotional activities are rare since RC are usually well known among their target group and do not have to compete with other airlines on the same routes.[57]

3.3.2 Advantages and Disadvantages

RC try to combine some of the FSNC's and some of the LCC's advantages in attempt to minimize the overall disadvantages. For example, they operate with less heterogeneous fleets than FSNC, but still have more than one or two aircraft types within their fleet. They are thus able to react to changes in demand, but can also keep maintenance costs at a lower level. Furthermore, RC approach to smaller airports with lower airport fees than FSNC, which allows them again to save costs. Additionally, RC can benefit from their mother companies reputation, which again saves them costs due to a less intensive promotion. Besides, they serve as feeder for their mother company's, which again saves costs due to a less intensive promotion. However, such a high dependency on the parent company also brings some risk along. RC rely on the mother company's performance as well as on its reputation. If the mother company fails, the RC is at risk of failing, too.

---

[56] Conrady, Fichert, and Sterzenbach (2013), p. 231-232.
[57] Conrady, Fichert, and Sterzenbach (2013), p. 231-233.

3.4 Leisure Carriers

3.4.1 General Characteristics

Leisure Carriers, formerly known as Charter Carriers, were once established to serve tour operators, which used to charter whole aircrafts in order to include the seats in their pre-packaged tours, which they then sold to their final customers for a packaged price. However, this product changed to a more individual one, the dynamic package. The customer now builds his or her own package. Therefore, a whole aircraft is not needed anymore. Tour operators are able to book seats separately. For this reason LC do not fly on demand only anymore, but on fixed schedules, which do depend on the tour operators' demand though.[58]

LC mainly operate continentally serving tourism destinations with point-to-point-connections. However, if the demand within a certain source market is not sufficient, LC are able to stop at other source markets to fill up the missing seats. Thereby, LC are able to react according to demand shifts more easily than other carriers. Additionally, they distinguish from other carriers by approaching to airports of any size.

Their fleet usually comprises a variety of different aircrafts providing a capacity ranging from 150 to 250 seats each. LC mainly target leisure travelers.[59]

3.4.2 Marketing Mix

LC usually offer one booking class on a medium level only. Services on board, such as snacks and the transport of a piece of luggage, are included in the final price. Prices vary depending on whether the flight is purchased as a component of a packaged tour by a tour operator or as a single item by a final customer. Tour operators usually buy a contingent of seats to individual contract conditions. Thus, prices vary. However, by receiving a special offer, the tour operator is responsible for selling all flights. The final customer, who buys one flight only, pays a low price, which is comparable to prices offered by LCC.

---

[58] Conrady, Fichert, and Sterzenbach (2013), p. 247 f; Pompl (2007), p. 35-38.
[59] Conrady, Fichert, and Sterzenbach (2013), p. 248 f; Pompl (2007), p.104.

As it is already mentioned above, LC mainly distribute their products and services to tour operators, which in return sell them to their final customer (e.g. via travel agencies). Additionally, they distribute them directly to the final customer via their own webpage or via their own call centers. As a consequence of selling the majority of their products to tour operators, LC only need to promote their products and services to their final customers, who buy them individually.[60]

It can be said that the core business of LC is still done in cooperation with tour operators. Prices and products are designed according to contracts and individual tour operator needs. Thus, LC are mentioned and explained at this point for the sake of completeness only and shall not be taken into further consideration.

3.5 Overlapping Business Concepts

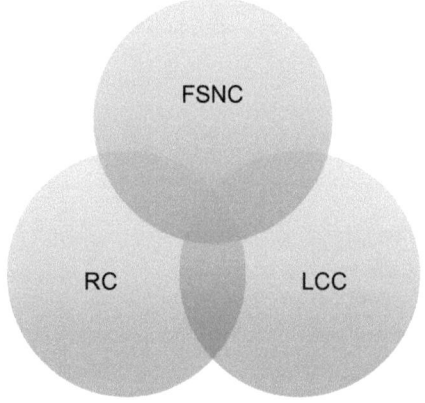

Figure 1: Overlapping Airline Business Concepts[61]

In order to achieve the best business performance, airlines constantly keep on renewing their marketing mix by designing new products, establishing new

---

[60] Conrady, Fichert, and Sterzenbach (2013), p. 249 f; Pompl (2007), p.104.
[61] Own illustration.

price concepts, developing new distribution channels and finding new promotion possibilities. This leads to the creation of new business concepts, which cannot be distinguished clearly as being either a FSNC, a RC or LCC due to an overlap of multiple criteria.

4 Analysis of the German Passenger Airline Market – Porter's Five Forces

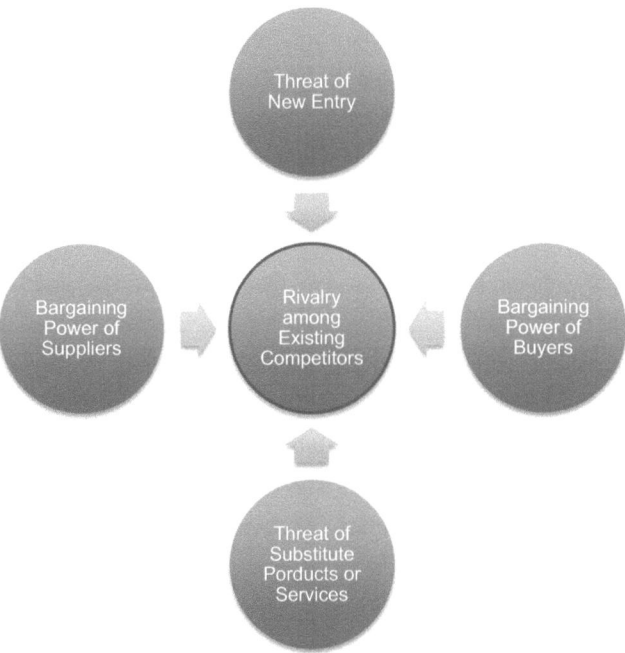

Figure 2: Michael Porter's Five Forces Model[62]

PORTER argues that five different forces drive a company's performance. These are on the one hand the direct competition, which is offering the same product (competitive rivalry), and on the other hand the indirect competition, with whom the company is fighting for profits (suppliers, customers, new entrants and substitutes). By looking at these five forces it can be exposed whether an industry is profitable, what is driving the profitability and what the trends that can lead to a change are.

---

[62] Michael Porter- On Five Forces Model (2011).

4.1 Rivalry among Existing Competitors

PORTER argues that the company competes with the direct competitors for profitability.[63] Direct competitors in the case of an airline, which sell single seats, are other airlines, which also sell single seats. These are Full Service Network Carriers, Low Cost Carriers and Regional Carriers. Thus, the following sub-chapter will have a look at the competition among these carrier types within the German passenger airline market. Although LC do not sell single seats and are, thus, not direct competitors, they are still mentioned in certain statistics in order to create a complete picture of the German passenger airline market.

The list of airlines operating on the German passenger market changes constantly due to bankruptcies, new start-up companies, as well as mergers and acquisitions. Nevertheless, the number of airlines usually adds up to approximately 150 different carriers. The biggest airlines according to their number of flights offered on the German market are Lufthansa, AirBerlin, Germanwings, Ryanair, Air France, easyJet, KLM-Royal Dutch Airlines, Swiss, SAS Scandinavian Airlines, British Airways, Condor, Austrian Airlines AG, TUIfly, Turkish Airlines, LOT-Polish Airlines, Aeroflot Russian Airlines, Cirrus Airlines, Luxair, SunExpress, Intersky, Flybe, Iberia, Emirates and Brussels Airlines.[64] Table 1 classifies the major airlines according to their business concepts.

---

[63] Michael Porter- On Five Forces Model (2011).
[64] Deutsches Zentrum für Luft- und Raumfahrt e.V. (2012).

Full Service Network Carriers

- Aeroflot Russian Airlines (Skyteam)
- Air France (Skyteam)
- Austrian Airlines AG (Star Alliance)
- British Airways (oneworld)
- Czech Airlines (Skyteam)
- Finnair (oneworld)
- Iberia (oneworld)
- KLM-Royal Dutch Airlines (Skyteam)
- LOT-Polish Airlines (Star Alliance)
- Luxair
- SAS Scandinavian Airlines (Star Alliance)
- Swiss (Star Alliance)
- Turkish Airways (Star Alliance)

Low Cost Carriers

- AirBerlin (oneworld)
- easyJet
- Flybe
- Germanwings
- Intersky
- Ryanair

Regional Carriers

- Cirrus Airlines
- OLT Ostfriesische Lufttransport
- Sylt Air

Leisure Carriers

- Condor Flugdienst
- Hahn Air
- Hamburg Airways
- Pegasus Airlines
- SunExpress
- Tuifly

Table 1: Classification of Airlines Operating in Germany[65]

---

[65] The airlines are assigned to the carrier of which they fulfill the most criteria. [Deutsches Zentrum für Luft- und Raumfahrt e.V. (2012)].

The majority of all flights is offered by FSNC and LCC (92,8 percent) as shown in figure 3. FSNC provide almost two thirds and LCC provide a quarter of all flights. LC and RC, in comparison, offer a small portion of about seven percent only. Once more, this again shows that these business concepts, even though they are also selling a number of single seats, are not playing a major role within the German passenger airline market.[66]

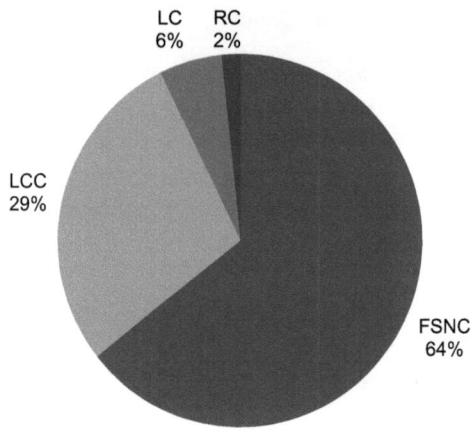

Figure 3: Market Share of Business Concept in Germany in 2011[67]

Both, FSNC and LCC, experienced a decline in their market share in 2010. Yet, in 2011 FSNC were able to exceed their level of 2009, whereas LCC were subject to a further downturn by 1,8 percent that year. More detailed information can be found in the appendix in table 7 and figure 25.

4.1.1 Competition among Low Cost Carriers

The number of routes offered by LCC on the German passenger airline market increased constantly by 35 percent on average per year until 2008. After that year, the growth rate declined. In 2009 as much as six percent of all routes were cancelled and not offered anymore. While in 2010 the level of 2008 was

---

[66] Own illustration on the basis of Deutsches Zentrum für Luft- und Raumfahrt e.V. (2012).
[67] Deutsches Zentrum für Luft- und Raumfahrt e.V. (2012).

reached again, the growth rate of the number of routes offered by LCC has never returned its peak level of the beginning of the millennium (figure 4).

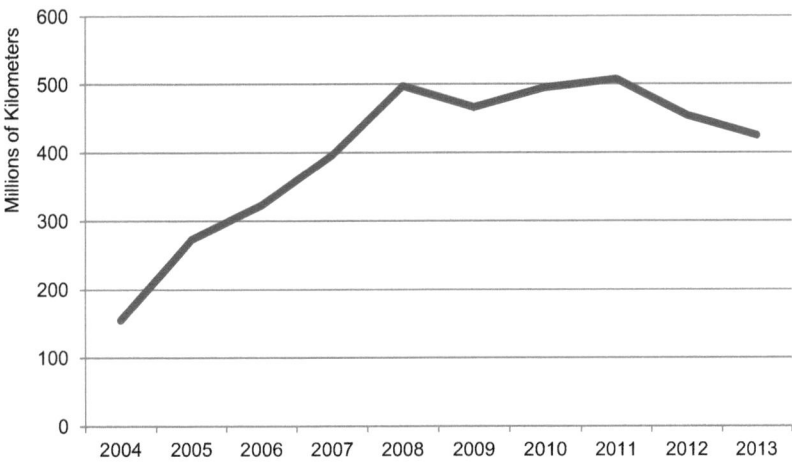

Figure 4: Route Network offered by LCC in Germany (January of each year)[68]

A total of 21 LCC operate on the German passenger airline market. The ten largest based on the number of flights offered are AirBerlin, Germanwings, Ryanair, easyJet, Intersky, Wizz, Are Lingus, Flybe, Norwegian and Air Baltic. Among theses, nearly, half of all low cost flights are provided by AirBerlin (47 percent), followed by Germanwings with 21 percent, Ryanair (ten percent) and easyJet (nine percent). Yet, there are 21 airlines termed as LCC; not all of them offer the classical concept, which is described in chapter 3.2. AirBerlin, for example, offers international flights and a frequent flyer program, which is not characteristic for an LCC. Additionally, AirBerlin joined a strategic alliance recently.[69] Thus, the airline is no LCC in the classical meaning anymore. However, due to some characteristics, such as low fares, it is still indicated as an LCC by most institutions.

---

[68] Own illustration on the basis of Deutsches Zentrum für Luft- und Raumfahrt e.V. and Arbeitsgemeinschaft Deutscher Verkehrsflughäfen (2007; 2010; 2013).
[69] AirBerlin (2013).

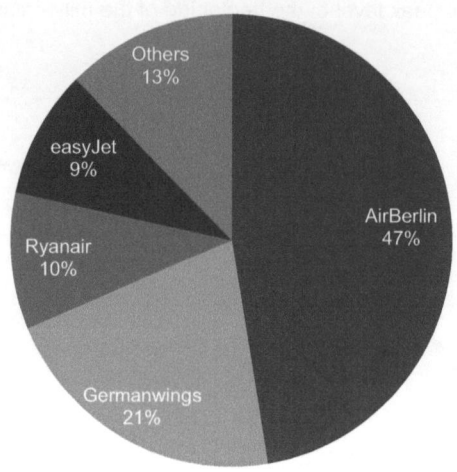

Figure 5: LCC Market Share in Germany in January 2013[70]

Even though there are several LCC operating on the German passenger airline market, competition among those exists on only four percent of all routes. The remaining 96 percent of routes served by LCC, are served by one single airline only.[71] This does not take into account flights from different airports within the same area to the same destination, which increases the competition again.

4.1.2 Full Service Network Carriers and Strategic Alliances

FSNC operated almost 65 percent of all flights departing on the German passenger airline market in 2012 (figure 3). The majority (75 percent) of these flights was carried out by strategic alliances, such as Star Alliance (35 percent), SkyTeam (25 percent) and oneworld (15 percent). More information about the performance of the different strategic alliances can be found in the appendix in table 8 and figure 26.

---

[70] Own illustration on the basis of Deutsches Zentrum für Luft- und Raumfahrt e.V. (2013).
[71] Deutsches Zentrum für Luft- und Raumfahrt e.V. (2013).

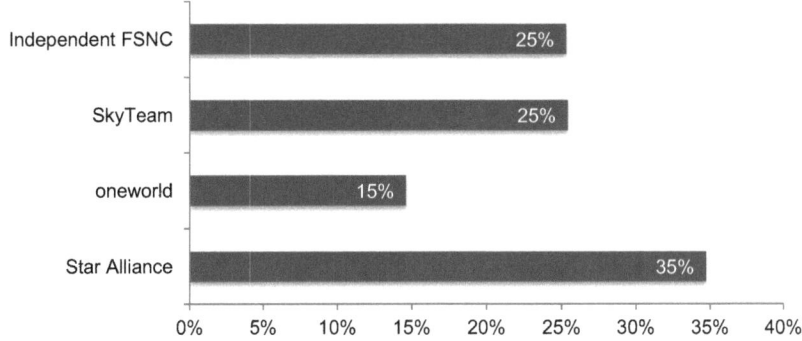

Figure 6: FSNC and Strategic Alliances Market Share in Germany in 2011[72]

4.1.3 Differentiation of Products and Services

Classical business concepts are overlapping in many cases. As mentioned in sub-chapter 4.1.1, AirBerlin offers services, which are not characteristic for LCC. In the same way, some FSNC are starting to exclude services in order to reduce costs and ticket fares in return. As a result, the products offered by FSNC and LCC start to appear more similar, which may increase the risk of travelers switching the airline quickly. Thus, airlines which used to sell different products to different clients may become competitors, which increases the overall competition in return.

4.1.4 The German Passenger Airline Market in Europe

With 25,6 billion dollars in market value, the German passenger airline market accounts for 14 percent of the total European airline passenger market (figure 7). Thus, Germany was the biggest passenger airline market in Europe in 2011. Such a position may attract airlines to enter the German market, which might result in a higher competition in return.[73] Further information on the regional share can be found in the appendix (table 6).

---

[72] Own illustration on the basis of Deutsches Zentrum für Luft- und Raumfahrt e.V. (2012).
[73] MarketLine (2012), p. 12.

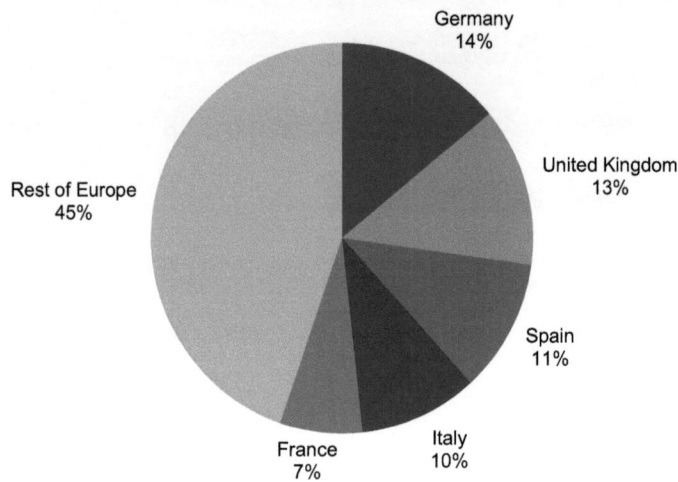

Figure 7: Geographical Segmentation of the European Airline Industry Value[74]

4.2 Threat of New Entry

The first of the forces, which may disturb the competitive equilibrium of the mentioned existing rivalry through indirect competition, is the threat if new entrants. The following sub-chapters will give an insight into the German market entry barriers in order to point out the likeliness of new airlines to enter the German passenger market.

4.2.1 Entry Barriers through Slot Allocation Regulations

The slot allocation at Germany's sixteen major airports happens via the Airport Coordination Germany, a non-profit organization which is financed by German airlines. The organization distinguishes between coordinated and facilitated airports. Coordinated airports are those that register a higher demand than slots they can offer. Accordingly, all airlines need to have a slot to arrive or to depart allocated by the country's slot coordinator. In 2013 the airports of Düsseldorf, Frankfurt/Main, Munich, Stuttgart as well as Berlin-Schönefeld and Berlin-Tegel belong to this category. At facilitated airports, in comparison, the

---

[74] Own illustration on the basis of Deutsches Zentrum für Luft- und Raumfahrt e.V. (2012).

demand of slots by airlines exceeds the supply only sometimes, e.g. on special weekdays. For this reason an airport coordinator is not required. Here, airlines have to cooperate with each other voluntarily. The airports of Bremen, Cologne, Dresden, Erfurt, Munster-Osnabruck, Hanover, Hamburg, Leipzig-Halle, Nuremberg and Saarbrucken belong to this category.[75] However, facilitated airports have to report to the Airport Coordination Germany as well. If they recognize a demand which cannot be handled by the airport and the airlines themselves anymore, they can be announced as a coordinated airport immediately. This is how they are distinguished from the remaining airports, such as Dortmund or Paderborn. At these airports the demand does not exceed the supply at any time. For this reason the airports as well as the airlines are responsible for the slot allocation at all times. They also do not have to report to the Airport Coordination Germany.[76]

At coordinated airports the slot allocation is done according to certain priorities by the council regulation (EEC) No 95/93 taking into account advice of the International Air Transport Association. Accordingly, slots are allocated to airlines that have used them in the previous period already to at least 80 percent. These rights are also known as "grandfather rights" or "80-20 rule".[77] 50 percent of the remaining slots are given to new applicants. The other half is primarily given away to airlines that offer scheduled flights all year around as well as periodic occasional flights, firstly. Then airlines which offer scheduled as well as periodic flights in certain seasons are taken into account. Afterwards, flights are allocated to airlines which provide occasional flights and lastly, non-commercial flights are considered.[78]

This way of slot allocation creates a barrier for airlines which want to enter the German passenger market. At coordinated airports the entry may even not be possible at all since these airports are mainly dominated by FSNC, which have been operating there for many years already. Consequently, the majority of slots available is given to them according to the "grandfather rights". New entrants who want to start their business with occasional flights only, thus, rely

---

[75] Airport Coordination Germany (2013).
[76] Centrum für Europäische Politik (2012).
[77] European Commission (2011).
[78] European Communities (1993).

on the bigger airlines. Such a dependency may also occur at facilitated airports, where new entrants have to compete with other airlines, which may be in a better negotiating position due to more experience and already existing relationships to other airlines.

Furthermore, it is being argued, that a number of slots cannot be used due to delays resulting from a complex air traffic control[79], which used to be done by each European country individually. Yet, EUROCONTROL, the European Organization for the Safety of Air Navigation, consisting of 39 members (European Union Commission and other states of the European area) has been creating the Single European Sky, the completion of which will still take some time.[80]

4.2.2 The Liberalization of European Air Traffic

While the European Community treaty already contained some information on air traffic policies, the implementation of it went not as it was supposed to. The European air traffic was still regulated by each country individually and some bilateral treaties until the seventies. Thus, a number of different treaties and regulations existed, which resulted in a complex and rather confusing system. To eliminate this weakness, the deregulation of the European air traffic was started in 1974 by the European Court of Justice, which remitted some general rules. Later on, to finally create one single aviation market among the European Community, three liberalization packages became effective, the first one in 1988, the second one in 1990, and the third one in 1993. All packages contained regulations, which enabled airlines to make their own decisions on market access, capacity and fares. Cross-border traffic has been almost deregulated ever since. Only inland traffic through suppliers of other European Union member states had not been possible without restrictions until 1. April 1997.[81]

In 2008 the liberalization packages were replaced by regulation (EC) 1008 / 2008 of the European parliament and the council. The regulation took over the

---

[79] Loppow (1997).
[80] EUROCONTROL (2013).
[81] Conrady, Fichert, and Sterzenbach (2013), p. 48-50.

main content of the liberalization packages, developed it further and added more information.[82] These regulations also apply for some countries, which are not part of the European Union, but of the European area (e.g. Switzerland, Norway, Iceland, Liechtenstein).[83]

Through the liberalization process some barriers to enter the German passenger airline market have been brought down. Still, some individuals, such as Karl-Heinz Neumeister, the chief executive officer of the Association of European Airlines, were skeptical whether the decentralization would lead to a higher competition due to the already mentioned lack of slots at coordinated airports.[84] His assumption turned out to be correct. Yet, airlines, such as Ryanair (registered in Ireland) and easyJet (registered in the United Kingdom), for example, have proven that airlines of countries, which belong to the single common aviation area, are able to enter the German market and to widen their operations within the country by mainly operating at regional airports, which are neither facilitated, nor coordinated, but attractive through low fees.

4.2.3 Subsidies at Regional Airports

At regional airports airlines benefit from subsidies given by the state and federal states. In many cases, these subsidies make the survival of the airport possible in the first place. According to some authors, a shortage or the complete removal of this financial support might result in the bankruptcy of some airports, which are not able to cover their operational costs themselves. Furthermore, these airports have to cope with a high level of competition. The airports of Kassel-Calden, Dortmund and Paderborn, for example, are located within a distance of approximately 150 kilometers of each other, and have, thus, the same catchment area. To attract airlines they need to keep their fees down, which in return is not possible without subsidies at public expenses (taxes).[85] However, these subsidies also lead to a distortion of the competition. LCC are attracted by low airport fees and can start their business easier than at

---

[82] Euopean Parliament and European Council (2008).
[83] Conrady, Fichert, and Sterzenbach (2013), p. 48-50.
[84] Loppow (1997).
[85] Streule (2012).

coordinated or facilitated airports. For this reason, these subsidies are criticized by many stakeholders, such as airlines, which operate at coordinated airports, like Lufthansa.[86] Thus, the European Union Commission is planning to take away this facilitation of market entry. The commission is planning to enact a law in 2014, which is intended to reduce subsidies iteratively according to the airport's passenger numbers.[87] Consequently, LCC will not be able to enter the German market via regional airports as easily as they used to anymore. It can be said that such a law reduces competitive advantages on the one hand, but creates market entry barriers on the other hand.

4.2.4 The Elimination of Further Subsidies via the Aviation Tax Act

In 2011 the German government enacted a law on air traffic tax that commits each passenger to pay a tax on each flight originating in Germany. The law distinguishes between short-haul (up to 2.500 km), medium-haul (up to 6.000 km), and long-haul (more than 6.000 km); currently each flight is taxed with 7,50 Euro, 23,43 Euro or 42,18 Euro respectively.[88] The objective of the Aviation Tax Act has been to eliminate the competitive tax advantage, which the airline industry had held until then, against all other modes of passenger transportation, which always had to pay a value added tax of 19 percent on the mentioned distances. Furthermore, air traffic is known for being the most climate-damaging mode of transportation. Thus, through the elimination of this subsidy, the government is attempting to reduce the total number of flights taking off in Germany in order to antagonize the climate change and to purge the air, to reduce the noise exposure and, in return, to improve the health level of all people.[89] Airlines on the other hand argue that the tax will not achieve the desired outcome. Lufthansa, for example, even asserted, that passengers would go to neighboring countries to depart and that those passengers would accept farther travel to their departure airport, which would result in an even higher damage of the environment.[90] Now, that the law has been in operation

---

[86] Lufthansa (2012b).
[87] Döring (2013).
[88] "Luftverkehrsteuergesetz" (2010).
[89] Verkehrsclub Deutschland e.V. (2012a).
[90] Lufthansa (2010b).

for three years, some organizations as well as some airlines try to analyze whether the law has achieved its purpose.

Again, opinions are very divergent. A German organization, the German alliance for environment and nature protection (BUND für Umwelt und Naturschutz Deutschland e.V.), for example, argues that there is no trend visible that German passengers have been travelling from neighboring countries ever since the law has been implemented[91], whereas some German Airlines seem to be convinced that this is the case.[92] It will probably take a longer period of time as well as more detailed analyses to be certain, why the numbers of passengers have changed the way they did (see chapter 5.2). However, it already can be said, that the Aviation Tax Act created an entry barrier for airlines to enter the German passenger market.

4.2.5 Competitive Disadvantages Through Emission Trading

Emission trading was implemented in 2005 in order to reduce CO_2-emissions and to protect the environment sustainably in return. The principle of emission trading is a cap of the amount of CO_2-emission per period allowed, which is set by the European Union. Each company needs a certificate per CO_2-tonne produced. If at the end of a period the amount of emission does not meet the number of certificates owned, the company has different options. In case of a deficit it may either purchase additional certificates from other companies or it may invest in activities to reduce its CO_2-emissions. In case of a surplus of certificates the company may sell its certificates to other companies. Emission trading is intended to motivate companies of the European region to reduce their CO_2-emissions and to invest in innovations that support this process.[93]

In 2008 the European Union decided to include air traffic into the emission trading as well, starting in 2012. This should have covered all flights starting and landing in Europe including flights operated by airlines outside the European Union. However, countries outside Europe protest since they have been willing to find a climate protection agreement on an international basis first. Thus, the European Union decided to exclude flights from and to Europe

---

[91] During (2012).
[92] BUND für Umwelt und Naturschutz Deutschland e.V. (2012), p. 2.
[93] Bundesministerium für Umwelt, Naturschutz und Reaktorsicherheit (2013).

from emission trading until fall 2013 in order to prevent international trading conflicts.

While this exception also includes flights offered by European airlines, these still feel disadvantaged since airlines from outside Europe are able to offer their flights outside Europe under better conditions.[94] Thus, they might be able to cover higher cost of flights taking place within Europe by higher profits, which they achieve with flights outside of Europe. In consequence, a barrier to enter the European, including the German passenger airline market was created. And other markets have become a bit more attractive to enter for airlines.

4.2.6 Relatively High Capital and Resource Requirements

In order to run a passenger airline, various production factors are necessary. CONRADY, FICHERT and STERZENBACH distinguish between five different groups. The fifth group is the external factor, which in this case is the traveler and will be discussed in chapter 5 in detail. Thus, only four of five groups shall be stated at this point of the paper:[95]

- Human resources, such as flight personnel (e.g. pilots) and ground handling personnel, comprising all employees.
- Operational resources in the sense of tangible assets, such as aircrafts and terminals at the airports, which can be used for a long term.
- Raw materials, such as electricity, fuel and water, which are consumed on a regular basis.
- Intangible production factors, such as rights and licenses, which are usually also purchased for a longer period of time.

To acquire each factor a certain capital is needed. Here existing airlines might have some advantages over new entrants, such as a good reputation to hire the best employees and to negotiate contracts, a budget to purchase a higher amount of aircrafts, which leads to a lower price per aircraft in return. While an existing airline might have an advantage, it might also have a disadvantage due

---

[94] Kotowski (2012).
[95] Conrady, Fichert, and Sterzenbach (2013), p. 138 f.

to a bad reputation, for example. Thus, it depends on the situation whether it is easy to enter the airline passenger market in general. Overall it can be said that the high budget which is needed, creates a barrier to enter the German market.

4.3 Threat of Substitute Products and Services

Substitution within the German passenger airline market may either occur through replacing the core product of travelling from destination A to destination B by different modes of transportation, or by offering services, which make a journey unnecessary in the first place. Both options will be discussed further within the next two sub-chapters.

4.3.1 Intermodal Competition by Road and Rail

The core product, which is the traffic performance, can be substituted intermodally by rail and road.[96] Which mode of transport will be chosen depends on a number of factors. Most relevant are the fare, the comfort provided, the number of stopovers, the level of flexibility as well as the reliability in the sense of punctuality (see sub-chapter 5.1.). Different authors argue that within a distance from 800 to up to 1.000 kilometers rail and road offer some advantages through which they are able to compete with air traffic.[97] These are, for instance, city-center-to-city-center connections, short ground handling processes, and low to very low fares. Additionally, trains offer an overall low journey time due to speed trains.[98]

The number of substitute products available has been growing during the last years caused by the liberalization processes of the European passenger rail traffic and the German long-haul traffic of scheduled buses.

It has already been the aim of the German rail reform of 1994 by the German government to provide all rail companies based within the country with equal rights for accessing the German rail network. However, due to a lack of

---

[96] Waterways are not mentioned, as the portion of people traveling via ship for the sake of traveling from A to B is very low.
[97] Conrady, Fichert, and Sterzenbach (2013), p. 119.
[98] Verkehrsclub Deutschland e.V. (2012b).

common technical standards, the entry of the German rail market was complicated for foreign rail companies. In 2008 these standards were determined by the European Community and have been implemented by the European Union member states since. Additionally, in 2010 the European rail network was opened for cross-bordering passenger traffic for all in the European Union registered rail companies. [99] The consequence of this liberalization process has been the emergence of several rail traffic suppliers operating in Germany and connecting German cities with cities of bordering countries within a reasonable time frame and for an affordable amount of money.

Additionally, the German long-haul traffic of scheduled buses has been decentralized. Since the beginning of 2013 bus companies have been allowed to offer scheduled long-haul connections within Germany. Previously, it was only allowed on routes coming from and going to Berlin in order to keep the competition or the German rail traffic low.[100] Due to this new law a number of different bus companies started their business connecting German cities for very low fares.[101]

Another option of travelling from A to B is by car. People can either rent a car from one of the car rentals or car sharing companies, or they can be a part of the non-commercial model of car sharing.

Consequently, it can be said that there is a high number of potential options of travelling within a distance of up to 800 to 1.000 kilometers, which form a threat for the German passenger airline market. Furthermore, the in sub-chapter 5.2.1 mentioned reduction in passenger numbers in 2013 might have been caused by these substitutes and not as argued by some stakeholders by the Aviation Tax Act (sub-chapter 4.2.4). However, when it comes to travelling to

---

[99] Bundesministerium für Verkehr, Bau und Stadtentwicklung (n.a.).
[100] Deutsche Presse Agentur (2012).
[101] Kamann (2013).

destinations which are further away, the threat of substitutes becomes relatively low due to the travelling time, which is on longer routes shorter by airplane.[102]

4.3.2 Substitution Through New Media

Some authors argue that new media offers opportunities, which make journeys unnecessary entirely. SHAW, for instance, says that the purpose of travelling for business travelers can also be met by video-conferencing, telephone-conferencing and email.[103] However, a glance at sub-chapter 5.1 shows that the aim of business travelers is not only to get in contact with business partners and to negotiate contracts. It is rather the personal contact that is important to the business traveler, which, in return, makes a journey necessary. Thus, media cannot always substitute business travelling in any case, not even during an economic recession.

Moreover, CONRADY, FICHERT and STERZENBACH mention the thesis of complementarity. It indicates that the technical progress has intensified the worldwide division of labor and has, thus, increased the need for traveling in order to get in personal contact with business partners.[104] This thesis confirms what business travelers say about their traveling purposes (sub-chapter 5.1).

Consequently, it can be said that there might be a number of business travelers who stop traveling because their traveling purposes can be fulfilled by the use of the media. Nonetheless, it can also be seen as a fact, that the number of business travelers will rise as CONRADY, FICHERT and STERZENBACH describe due to an overall business globalization. Thus, the decrease on the one hand will probably be balanced by the increase on the other hand and the emergence of new media cannot exclusively be seen as a threat of substitution for the German passenger airline market.

---

[102] Conrady, Fichert, and Sterzenbach (2013), p. 119.
[103] Shaw (2011), p. 86 f.
[104] Conrady, Fichert, and Sterzenbach (2013), p. 122.

4.4 Bargaining Power of Suppliers

PORTER argues that companies which mainly depend on one supplier only, are loosing profits due to the bargaining power of the supplier.[105] The following sub-chapters will look at different suppliers within the German passenger airline market, which are having a major impact on the airlines' profitability.

4.4.1 The Power of Aircraft Manufacturers

Airlines depend on aircrafts. Accordingly, they depend on aircraft manufacturers as well. The market share of aircraft manufacturers varies according to the aircraft's size and the distance, which has to be flown. The market for commercial passenger aircrafts can be divided into wide-bodied aircraft manufacturers, short to medium-haul aircraft manufacturers and regional aircraft manufacturers.[106]

In 1970 Boeing launched its wide-body aircraft 747 with an average capacity of about 400 seats.[107] Boeing used to be the only supplier of aircrafts of this size for more than 25 years. Thus, they held a monopoly position in the world's aircraft manufacturing market for wide-bodied aircrafts. However, when Airbus introduced its A380 in 2005 with an average capacity of approximately 500 seats[108], Boeing lost its monopoly position, and a duopoly was created and has existed ever since. Today some experts argue that this situation is subject to change again as the demand for the 747 is dropping. Thus, they expect Airbus to take over the monopoly position in supplying wide-bodied aircrafts.[109]

The same duopoly of Airbus and Boeing exists among the suppliers of aircrafts for short and medium-hauls. But, similarly, in this market experts expect a change to happen due to the market entry of Comac, a Chinese aircraft supplier, who introduced a competitive aircraft scheduled to go into service in

---

[105] Michael Porter- On Five Forces Model (2011).
[106] Conrady, Fichert, and Sterzenbach (2013), p. 144.
[107] Boeing (1995).
[108] AIRBUS (2013).
[109] n.a. (2013).

2016.[110] Thus, the duopoly would become an oligopoly, which would weaken the bargaining power of Airbus and Boeing accordingly.

The number of regional aircraft manufacturers is even larger. Here, the major suppliers are Airbus, Boeing, Bombardier and Embraer.[111]

Looking at the number of suppliers, it can be said that the smaller the aircraft, the greater the competition of suppliers and, hence, the lower their bargaining power.

4.4.2 The Bargaining Power of Airports

The situation described in the previous sub-chapter 4.4.1, can also be applied to airports. The smaller the airport, the smaller is its bargaining power.

The number of primary airports in Germany is low. The Frankfurt/Main and the Munich airport, for instance, serve as hub and register a high demand. Furthermore, they are located in two major cities of the country and provide a good infrastructure, which makes them unique. Consequently, they have a high bargaining power and can charge high fees, which in return reduces the profitability of airlines.

Smaller airports, which register a lower demand and less passengers, such as the Cologne/Bonn or the Hamburg airport have comparatively less bargaining power due to a higher number of suppliers available. However, since they are usually also located in big cities and provide a good infrastructure, they still have a considerably bargaining power.

The number of German regional airports, in comparison, is even higher and so is the competition among them. They have a lower or even a very low bargaining power. They have to keep their fees at a low level to attract the airlines in order to survive. Here the airlines have a better bargaining power since they can switch to other airports easily.[112]

---

[110] Hegmann (2013)
[111] Conrady, Fichert, and Sterzenbach (2013), p. 144.
[112] Spaeth (2012).

4.4.3 Global Distribution Systems

As mentioned in sub-chapter 2.2.3 many airlines distribute their products through GDS in order to reach a high number of customers and to save distribution costs. The major GDS are Amadeus, Sabre and Travelport.[113] There are further systems, which, however, do not play a significant role within the airline passenger market. According to the small number of three, the system providers are in a good negotiating position. Airlines depend on them if they want to sell their products through them. Consequently, GDS suppliers may reduce the airlines' profitability on the one hand. On the other hand, as mentioned before, through selling their products via a GDS, airlines are able to save distribution costs. Furthermore, GDS serve as marketplaces where success is higher if the number of customers and the number of service providers is higher.[114] Thus, a small number of GDS with a high number of customers and service providers respectively, is more successful than a high number of GDS with a low number of customers and service providers. Consequently, a low number of GDS reduces the airlines' negotiating power, but also offers advantages, which might lead to a higher profitability in return.

The list of airline suppliers could, of course, be continued. However, to get an overview of the negotiating position of airlines, the three major suppliers, aircraft manufacturers, airports and GDS providers shall be sufficient for the sake of this paper. They have demonstrated that there are usually one to three market leaders, which have a strong bargaining power. Negotiating with them might result in a low profitability. However, apart from these market leaders, there is usually a high number of smaller options. They have a lower bargaining power and, consequently, do not reduce the airlines' profitability.

4.5 Bargaining Power of Buyers

The last of Porter's five forces which has an influence on the profitability of a company, is the bargaining power of the buyer. PORTER argues that the number of buyers and the costs of switching the airline have an influence on

---

[113] Schulz (2010), p. 267.
[114] Conrady, Fichert, and Sterzenbach (2013), p. 479.

the airlines profitability.[115] The following sub-chapter will discuss these two variables, whereas the buyer is synonymous with the traveler.

4.5.1 The Number of Buyers

The number of buyers on the German passenger airline market has been growing during the last ten years (figure 13). This development has been caused by a number of different factors. The Gross Domestic Product (GDP), for example, has had a positive growth rate for the last 20 years except for the years 1993, 2003, and 2009 (figure 24). It indicates a steadily growing wealth of the society, which enables people to spend more money on travel. Furthermore, the emergence of LCC enabled less wealthy people to travel by airplane due to low fares as well.

Consequently, an airline does not depend on one buyer only, which reduces the bargaining power of the buyer as a result.

4.5.2 Switching Costs

However, even though the number of customers is high, the bargaining power of customers is not as low as it could be due to low costs for switching the airline. Private travelers, for example, are very price sensitive as shown in figure 11. Thus, they tend to compare prices before buying. The emergence of different comparison platforms on the internet, such as www.fluege.de or www.swoodoo.de has simplified the comparison of prices. Thus, private travelers can compare prices in a short period of time. Furthermore, LCC promote their low fares very aggressively. Consequently, travelers are aware of the existence of low fares and do not need to seek for them for a long time. These reductions in time can be seen as a reduction of the switching costs, which leads to an increase in the bargaining power of the buyer in return.

Additionally, substitute products are easy to access and easy to compare as well. Long-distance buses, for example, promote their products aggressively. Thus, the buyers can stay informed about the cheapest travel option available at any time and can switch to them easily.

---

[115] Michael Porter- On Five Forces Model (2011).

FSNC try to bind their buyers through frequent flyer programs, whereas the loss of points or miles through switching to another airline can be seen as switching costs. However, these customer loyalty programs become attractive when flying frequently, only.

Consequently, it can be said, that the bargaining power of the buyer through their high number gets balanced trough low switching costs. The bargaining power of the buyer within the German passenger airline market can, thus, be seen as moderate.

4.6 Analysis of the Results of Porter's Five Forces

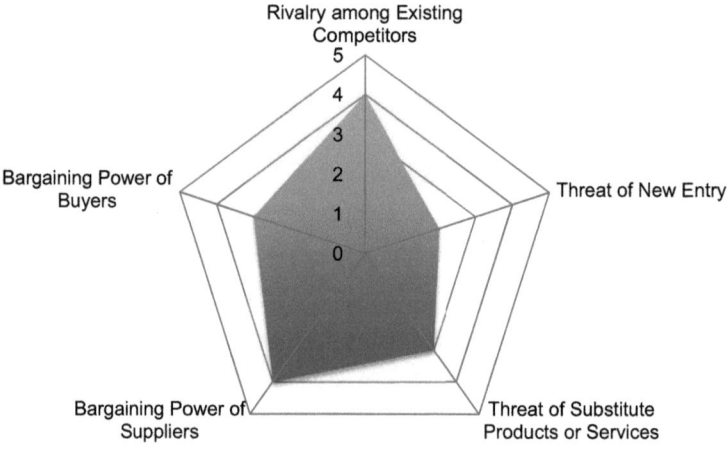

Figure 8:   Attractiveness of the German Passenger Airline Market[116]

It can be said that due to a high number of passengers and a positive market growth, a high number of airlines can be justified. However, the existing airlines seem to compete mainly via the price, but very little on the differentiation of

---

[116] 0=not existent 1=very weak, 2=weak, 3=neutral, 4=strong, 5=very strong (own illustration).

their products and services. Thus, the overall competition increases again, which leads to a strong rivalry among the existing competitors.

Although at coordinated and facilitated airports slots are given to existing airlines more easily than to new ones, new airlines have a good bargaining power at non-coordinated airports, such as Paderborn or Dortmund. Thus, also small airlines are given a chance to start operating in Germany. However, these airports exist through subsidies by the state, which are supposed to be reduced during the upcoming years. Consequently, new airlines will have to pay higher fees or will shift to other airports, where the slot allocation is regulated. As a result, new airlines will not be advantaged through lower fees anymore and a market entry barrier will be created. Additionally, the Aviation Tax Act and emission trading were implemented, which reduced the advantages the airline industry had over competing modes of transportation on the one hand, but created a market entry barrier for airlines on the other hand.
Consequently, it can be said, that the German passenger airline market is hard to enter due to the aforementioned factors.

The emergence of long-haul buses as well as new rail companies connecting city-centers has lead to a higher competition amongst the different modes of transportation upon a distance of 800 to 1.000 kilometers. However, the longer the distance which has to be travelled, the lower the threat of substitution by intermodal competitors becomes. Thus, the overall threat of substitution by products which fulfill the same needs is moderate. The same can be said about the threat of substitution by new media, through which a journey is not necessary anymore. It is compensated by a growing number of travelers, which are a result of the globalization process.

The list of alternative suppliers for one item is short, as it is explained in sub-chapter 4.4, which increases their bargaining power to a high level and lowers the bargaining power of the airlines in return.

The bargaining power of the buyer is balanced due to a high number of buyers on the one side and low costs of switching the airline on the other side.

According to these findings, it can be said, that the German passenger airline market is not a very attractive one to operate in for any airline.

5 The Demand within the German Passenger Airline Market

5.1 Business and Private Traveler in the Decision-Making Process

In the airline passenger market the customer and the consumer are not always the same person like they might be in other industries. In the airline passenger market the customer is the person who makes the decision and buys the ticket, and the consumer is the person who finally travels, but is not involved in the purchasing process of the ticket. However, for simplification reasons the following pages will not consider a differentiation between these two persons and will, thus, talk about 'the traveler' in general.

According to SHAW the general decision-making process of booking a flight can be divided into four different decisions which have to be made. In the first place it has to be decided whether the journey will be made at all. Secondly, a mode of transportation has to be chosen. Then, if the decision was made for air transport, a product (class, services, etc.) has to be selected. And finally, an airline has to be chosen.[117]

While all these decisions have to be made in order to book a flight, it can be argued whether this will always happen in the mentioned sequence. According to different preferences the third and the fourth decision might change their positions. In respect of their priorities they can, thus, be put in the same stage of the decision-making process.

Additionally, since this paper focuses on traveling rather than not traveling, the first decision question will not be considered at all. Consequently, only the following two stages within the decision-making process will be taken into account further in this paper:

1. Which mode of transport will be chosen?
2. Which airline and which product will be selected?

---

[117] Shaw (2011), p. 12 f.

Literature discusses mainly two types of travel within the passenger airline market.[118] DOGANIS, in comparison, argues that a segmentation of the market into two groups is not sufficient in order to attract the traveler appropriately. He, thus, distinguishes between four smaller sub-groups, whereas all passengers belonging to one segment are more homogeneous within their segment. Thus, he distinguishes between emergency business travel, routine business travel, two-week vacation and weekend trip.[119] A German organization, in comparison, the Gesellschaft für Konsumforschung SE Panel Services Deutschland, did a survey on the traveling behavior of private persons among 20.000 private households (approximately 45.000 people). Accordingly, they sub-divide the private travelers into five different groups as shown in table 2. However, this segmentation mainly focuses on journeys of private travelers, but does not consider all purposes of travelling. For example, people travelling to visit their family and friends or for business reasons are not considered. Thus, it can be said that all kinds of segmentations exist to divide the private travelers, but none of them seems complete or rather considers all possible reasons for travelling. Nonetheless, when deciding for a mode of transportation and choosing an airline, the decision making-criteria is about the same for all private and business passengers, but of different importance. Thus, this paper will only take into account the private and the business traveler.

---

[118] Conrady, Fichert, and Sterzenbach (2013), p. 110-116.
[119] Doganis (2010), p. 186-188.

Profile	Characteristics
Young Urban Hopper	• young people with little income (e.g. students) • book spontaneously • travel in spring and winter
Leisure & Family Tourist	• mostly families, between 30 and 44 years with average income • book some month in advance • travelling during summer for two weeks
Mediterranean Best-Ager	• older people (60+) wih average income • comfort matters • travel during fall for 8-14 days
Culture & Knowledge Seeker	• age 45+ • average income • more than one journey per year
Silver Traveller	• older people (60+) • high income • prefer longer journeys

Table 2: Profiles of Private Travelers[120]

The German organization Verkehrsclub Deutschland e.V. did a research on mobility needs of private travelers through interviewing 2.600 rail travelers. Figure 9 shows that flexibility matters for almost 75 percent of all private travelers. The components time and punctuality are important to half of the participants, whereas the price matters to 42 percent, only. Sustainability is of less importance for the private traveler.

---

[120] Own illustration on the basis of Initiative Airport Media and Gesellschaft für Konsumforschung (2011).

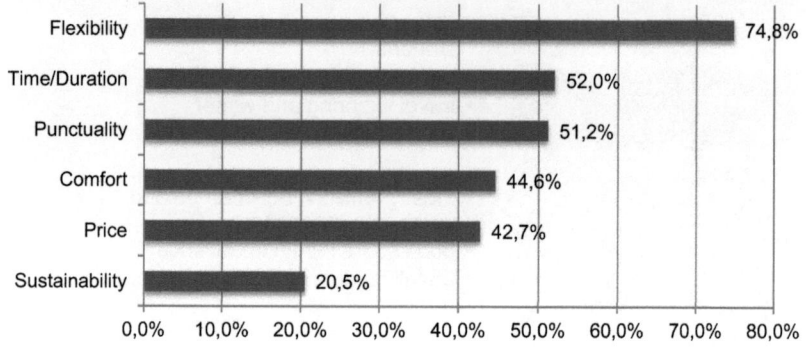

Figure 9:   Criteria for Transport Mode Selection[121]

However, the survey does not consider the distance of the total trip, which might also have an influence on the decision-making. Thus, the results demonstrate the general criteria for making a decision on the mode of transport, only.

The business traveler, in comparison, travels for business reasons, such as meetings, negotiations or fair visits. A German organization, the Deutsche Reiseverband e.V., which lobbies the interests of all German tourism companies, did a survey on business travel in 2013 in Germany. 200 managers of four different branches (service sector, industry, trade, consultancy) were asked about their preferences and habits when going on a business trip. The results show, that the main reasons for business journeys are to get to know business partners better, to maintain contacts to employees in branch offices and/or daughter companies, and to get in personal contact with business partners in general. Furthermore, as shown in figure 10, the participants were asked about the criteria which matter most to them, when booking a business trip. Almost 90 percent of all participants stated that they mostly or even always care about the fastest connection and 67 percent said that comfort matters to them. A low price and sustainability, on the contrary, are of less importance.[122]

---

[121] Own illustration on the basis of Verkehrsclub Deutschland e.V. (2009), p. 9.
[122] Deutscher Reiseverband e.V. (2013), p. 22.

This might be caused by the fact that the traveler is not spending his or her own but rather the company's money. Thus, the personal well-being can be prioritized when deciding on a mode of transportation.

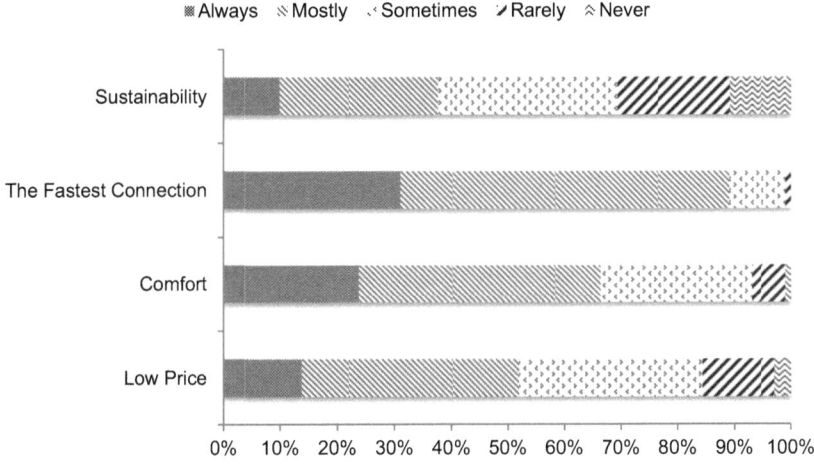

Figure 10: Decision-Making Criteria for Booking a Business Trip[123]

According to different authors, the list of criteria for both business travel and private travel gets smaller as soon as a mode of transportation is selected.

CONRADY, FICHERT and STERZENBACH distinguish between four main criteria which are considered when booking a flight. These are the price, flexibility, comfort and punctuality.[124] To name the price as a criteria and stating that it is of low importance for business traveler seems odd since it came up recently that also business travelers have to keep in mind their companies travelling standards, which in many cases includes a given budget. Thus, it can be assumed that the price in general does matter, but it does not necessarily have to be low. This criterion is also mentioned by DOGANIS, who also adds three more criteria to the decision-making process. However, he considers

---

[123] Survey Question: „Worauf achten Sie bei der Buchung Ihrer Geschäftsreisen?", N = 200. [On the basis of Deutscher Reiseverband e.V. (2013), p. 22.].
[124] Conrady, Fichert, and Sterzenbach (2013), p. 112-115.

more than the two mentioned target groups. Thus, his classification comprises decision-making criteria which are not necessarily considered by all travelers. For the sake of this paper, a differentiation between five criteria appears appropriate as shown in figure 11.

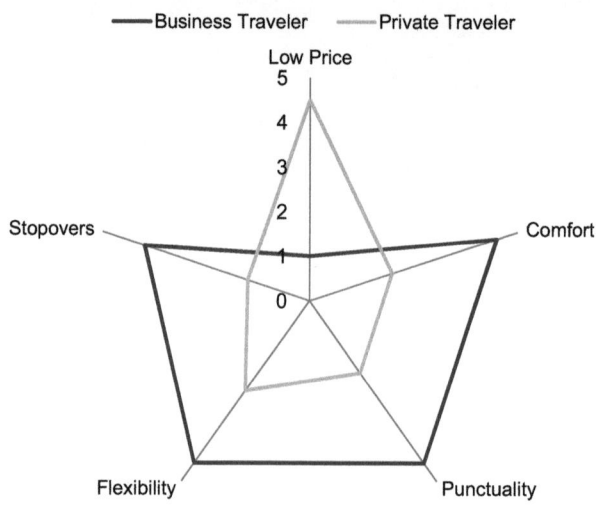

Figure 11: Decision-Making Criteria according to Preferences[125]

5.2 Development of the Passenger Demand in Germany

5.2.1 Development from 2001 to 2013 and Forecast

The total number of airline passengers departing and landing at German airports increased from 139,3 million in 2001 to 200,2 million in 2012 with an annual increase of approximately 2,4 percent on average.[126] As figure 12 shows, the increase in demand was reduced in some years due to different events, such as the terrorist attack in New York in 2001, the start of the world economic crisis in 2008 and the volcanic eruption in Iceland in 2010. Despite

---

[125] 0=not relevant 1=almost not relevant, 2=low relevance, 3=relevant, 4=very relevant, 5= highly relevant; (own illustration).
[126] Statistisches Bundesamt (2013a).

some setbacks, the airline passenger market always recovered to a certain degree. People decided to take off again and the growth in demand always came back to an annual average rate of approximately 2,4 percent.[127]

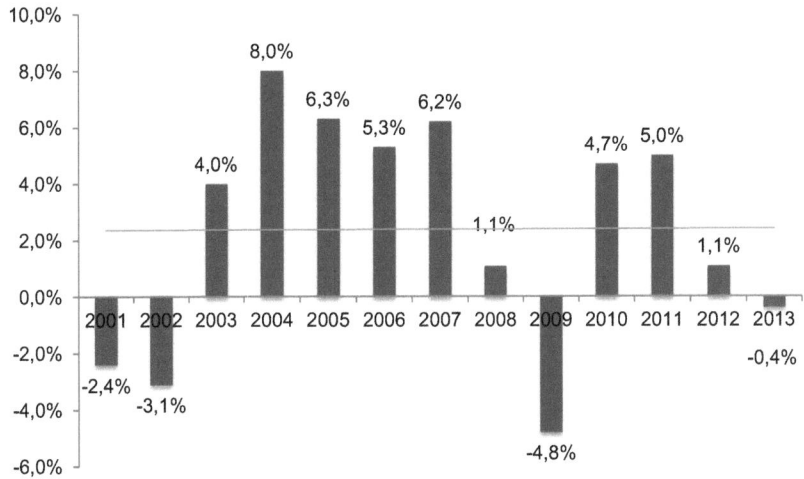

Figure 12: Annual Percentage Change of Passenger Volume (Arrivals and Departure) of Schedule and Charter Flights at Germany's 22 most Congested Airports[128]

Whereas figure 12 illustrates the change in demand at German airports for traffic (arrivals and departures) within Germany, the European region and Non-European countries, figure 13, distinguishes between these three areas.

---

[127] Deutsche Flugsicherung (2013), p. 4.
[128] Own illustration on the basis of Arbeitsgemeinschaft Deutscher Verkehrsflughäfen (2002-2013b).

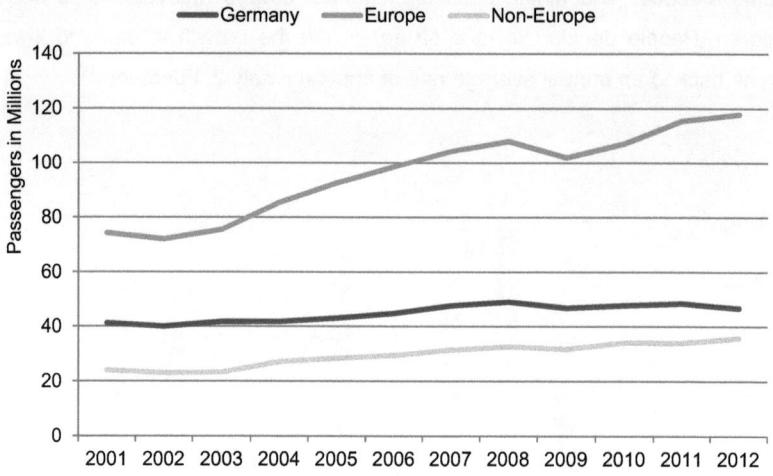

Figure 13: Passenger Volume at German Airports According to Destination and Origin Respectively[129]

Although all three areas show a steady increase in the volume of passengers departing and landing at German airports, the portions vary amongst each other. Passenger numbers within Germany increased from 2001 to 2012 by 1,3 percent on average with its peak of 6,3 percent in 2007. In contrast, airline passenger traffic between Germany and the European region increased by 4,9 percent and between Germany and the rest of the world increased by 5,2 percent on average. In both cases the highest growth rate was registered in 2004 (Europe with 13 percent and Non-Europe with 16,6 percent). Furthermore, in 2012 the number of passengers coming to and leaving Germany rose, but the number of people going by airplane within the country dropped by almost five percent compared to the previous year.

During the first half of 2013 passenger numbers at German airports dropped by 0,4 percent compared to the first half of 2012. The biggest reduction hit the airline passenger traffic within Germany with a minus of 5,9 percent. Even though the numbers looked better in the second quarter of the year, experts are

---

[129] Own illustration on the basis of Arbeitsgemeinschaft Deutscher Verkehrsflughäfen (2002-2013a).

skeptical whether the demand will reach the level of 2012 in the second half of 2013. They even doubt that they will go up before 2014 again at all. Although, the German passenger airline market realizes a negative growth rate at this time, experts expect the numbers of passengers to come back to a positive growth soon.[130] The aircraft manufacturer Airbus, in particular, expects a growth rate of 4,1 percent of the European airline passenger market and of 4,7 percent of the air passenger volume worldwide.[131]

5.2.2 The Development of Business and Private Travel

Figures 14 and figure 15 illustrate the development of the portions of private and business travelers from 2003 to 2012 using the example of the airport Cologne/Bonn. The percentage of business travelers dropped by seven percent during the last ten years, whereas private travelers increased their share from 59 to 66 percent.[132] As a consequence of the world economic crisis starting in 2008, German companies and organizations started to change their travel management[133], which might have caused the slight decrease in the portion of business travelers since 2008. Yet, it seems that this portion has started to recover again since 2010. The share of business travelers has been growing ever since. Additionally, the number of private travelers has been declining since 2010, which has also an effect on the total share.

The Verband Deutsches Reisemanagement e.V. interviewed 800 Germans who have been in charge of business travels within their companies about the German business travel market. One of the questions asked the interviewee about his or her prediction of the change of the number of business travels done in his or her company. More than half of the participants (57 percent) expect the number to stay the same as the previous year, whereas 29 percent expect the number to increase. Merely eight percent of the interviewees predict a reduction in the number of business travels.[134] If the participants are correct, it looks like the proportion of business and private travelers is on its way back to the level of 2003.

---

[130] Deutsche Flugsicherung (2013), p. 8.
[131] AIRBUS (2012), p. 85.
[132] Flughafen Köln Bonn GmbH (n.a.).
[133] Verband Deutsches Reisemanagement e.V. (2009), p. 17.
[134] Verband Deutsches Reisemanagement e.V. (2013), p. 19.

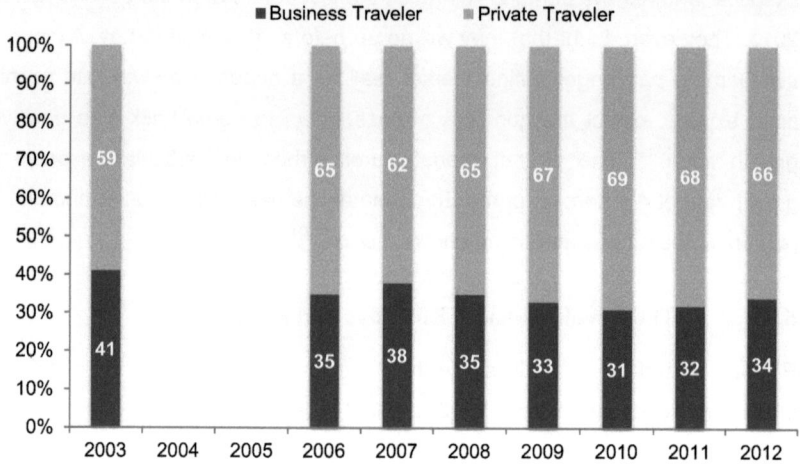

Figure 14: Portions of Business and Private Travelers at the Cologne-Bonn Airport [135]

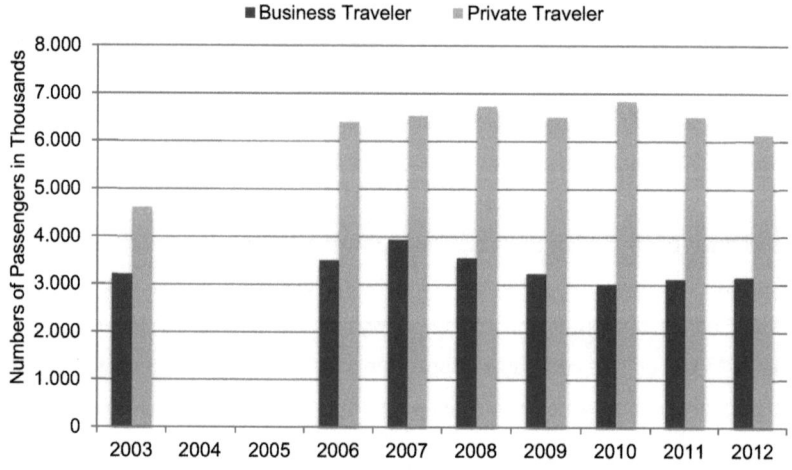

Figure 15: Total Numbers of Business and Private Travelers at the Cologne-Bonn Airport[136]

---

[135] Own illustration on the basis of Flughafen Köln Bonn GmbH (n.a.).
[136] Own illustration on the basis of Flughafen Köln Bonn GmbH (n.a.).

6   The Case of Lufthansa and Germanwings

The previous chapter gave an overview of the recent situation of the German passenger airline market and its challenges. In order to tackle these trials, airlines come up with different business strategies constantly. The German airline Lufthansa modified the business concept of its subsidiary Germanwings in 2013. Reasons for this change and the new strategy are discussed in the following chapter in order to evaluate whether the company has chosen a potential strategy to stay profitable and competitive.

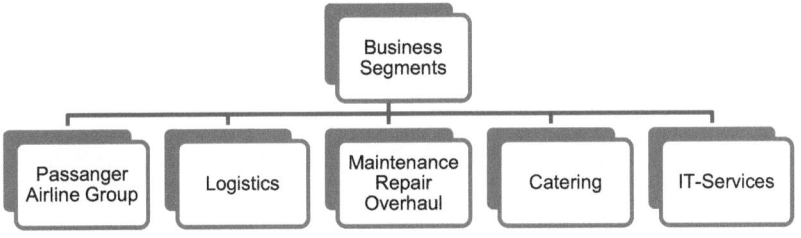

Table 3:   Business Segments of The Lufthansa Group[137]

6.1   The Lufthansa Group

The Lufthansa Group is comprised of five different business segments (Table 3), the Passenger Airline Group, which will be discussed in detail later on in this paper, Logistics (Lufthansa Cargo AG), Maintenance Repair Overhaul (Lufthansa Technik AG), Catering (LSG Lufthansa Service Holding AG) and IT-Services (Lufthansa Systems AG).

Additionally, Lufthansa runs its own provider of training courses for airlines, the Lufthansa Flight Training GmbH, and its own provider of Business Travel Management solutions, Lufthansa AirPlus.

---

[137] Lufthansa (2013a).

The Lufthansa Group mentions four main objectives on which all companies' activities are based on. These are the increase of the company value, the expansion of the market position of their airlines and service companies by actively designing the industry, the improvement of the customer satisfaction, and an economical and ecological sustainable business.[138]

Since all subsidiary companies belong to the airline industry, the profitability of the different segments may compensate for each other. However, belonging to one industry also means that all companies may suffer if the industry suffers. Hence, the stated mixture of companies does not only offer opportunities but also some risks.

The profits from all business segments' operations have dropped continuously from 1.378 million Euros in 2007 to 798 million Euros in 2012 (almost 60 percent). At the same time the share of all business segments has changed significantly. While in 2007 the Passenger Airline Group's operating result amounted to almost 60 percent of the Lufthansa Group's total operating result, the share has dropped to only 30 percent in 2012.[139] More information about the Lufthansa Group's operating results can be found in the appendix (figure 27).

This paper focuses on the German passenger airline market. Consequently, the following pages will only examine the business segment of the Passenger Airline Group of the Lufthansa Group further.

6.2   The Lufthansa Passenger Airline Group and its Business Performance

The Lufthansa Passenger Airline Group is the core business of the Lufthansa Group. It contains the Lufthansa Passenger Airlines including Germanwings, SWISS, Austrian Airlines and equity investments in Brussels Airlines, JetBlue and SunExpresss. It connects 250 destinations in more than 100 countries

---

[138] Lufthansa (2013b).
[139] Lufthansa (2008; 2009; 2010a; 2011; 2012a; 2013a).

through a global route network and slots at important hubs, such as Frankfurt, Munich, Zurich, Vienna and Brussels.[140] As a part of the Lufthansa Group it does not depend on external suppliers when it comes to catering, IT and Maintenance Repair Overhaul. Additionally, The Lufthansa Passenger Airline Group is able to train its own crew through the Lufthansa Flight Training GmbH. Some airlines of the Lufthansa Passenger Airline Group are a member of the strategic alliance Star Alliance, which enables them to achieve certain synergetic effects. These are, for instance, the common purchase of fuel and aircrafts, a global route network through code sharing[141], and a higher profit through economies of scale. These airlines are Austrian Airlines, Brussels Airlines, The Lufthansa Passenger Airline, and SWISS.

A further alliance is Lufthansa Regional. Four different airlines are operating under this brand. These are Air Dolomiti, Augsburg Airways, CityLine and Eurowings. Lufthansa Regional offers point-to-point flights across Europe and connecting flights to the Lufthansa hubs. Thus, it connects regional airports with the global network of Star Alliance.

Additionally, The Lufthansa Passenger Airline Group has a bilateral corporation with Air Malta, Air India, Jet Airways and Luxair. These corporations cover, for instance, the mutual recognition of each other's frequent flyer program, code sharing, and the adherence of common service and product standards to achieve customer satisfaction.
The Lufthansa Passenger Airline Group also is a part of two joint ventures through which the companies contribute production and management resources. These are the A++ transatlantic joint venture with United Airlines and Air Canada and the J+ bilateral Europe/Japan joint venture with All Nippon Airways.[142]

---

[140] Lufthansa (2013a).
[141] The term 'code sharing' is used to describe the method of an airline selling seats with their own flight number of a flight operated by a partner airline, the operating airline, which uses its own flight number as well. Consequently, it is not the code, which is shared, but the aircraft. [Conrady, Fichert, and Sterzenbach (2013), p. 275].
[142] Lufthansa (2013b).

Since 2007 the composition of the Lufthansa Passage Airline Group has changed various times through the purchase and the sale of different airlines, new partnerships, etc. Thus, the group has always managed to increase its passenger numbers. These have risen by 61 percent, from 62.878 thousand in 2007 to 103.051 thousand in 2012 (appendix, figure 29). At the same time the number of Available Seat-Kilometers (ASK) and the number of Revenue Passenger-Kilometers (RPK) have increased to about the same extend (65 percent, figure 16). Thus, the load factor[143] has always been at a level of approximately 78 percent (appendix, figure 29 and table 9).

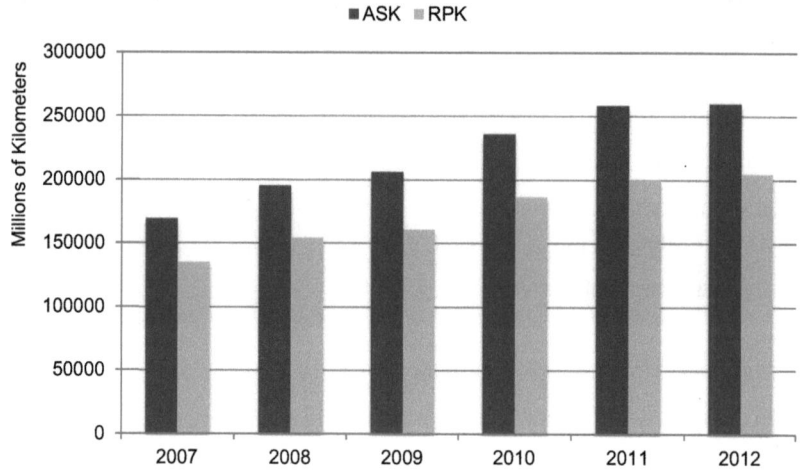

Figure 16:   Lufthansa Passenger Airline Group - ASK and RPK from 2007 to 2012[144]

The development of the group's operating result, in comparison, has been contrary (figure 17). It has dropped constantly during the previous six years from 826 million Euros in 2007 to 258 million Euros in 2012. The biggest breakin was registered in 2009. That year the Lufthansa Passenger Airline Group realized an operating loss of eight million Euros. Reasons for this are, on the one hand, a reduced demand as stated in sub-chapter 5.2, and growing expenses on the other hand (appendix, table 10).

---

[143] Load factor is the degree of capacity utilization.
[144] Own illustration on the basis of Lufthansa (2008; 2009; 2010a; 2011; 2012a; 2013a).

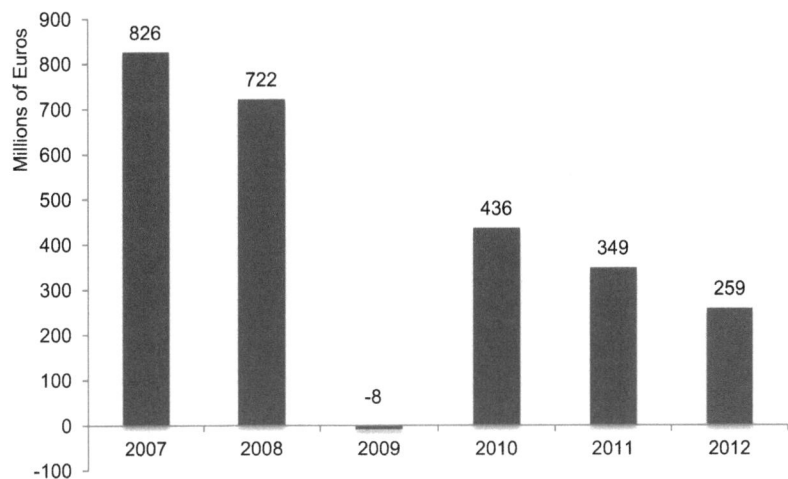

Figure 17: Lufthansa Passenger Airline Group - Operating Result from 2007 to 2012[145]

The Group also states that an increase of competition by LCC on flights operated on point-to-point routes beyond the hub airports of Frankfurt/Main and Munich is responsible for the declining operating result of the Lufthansa Passage Airline Group in general and flights operated under the brand of Lufthansa in particular. According to the Group the airlines AirBerlin and easyJet are the biggest competitors.[146]

It is arguable whether this is the case as it is stated in sub-chapter 4.1.1 that competition on routes operated by LCC only exists on four percent.

6.3 Business Performance of the Lufthansa Passenger Airline

The Full Service Network Carrier (sub-chapter 3.1) Lufthansa Passage conducts Lufthansa long-haul and medium-haul flights, as well as all flights operated by Lufthansa Regional. Germanwings was operated as a subsidiary company separately until 2011, but joined the Lufthansa Passenger Airline fully in 2012, yet it became a 100 percent subsidiary of the Lufthansa Group in 2009

---

[145] Own illustration on the basis of Lufthansa (2008; 2009; 2010a; 2011; 2012a; 2013a).
[146] Lufthansa (2012a).

already. For this reason performance numbers for both, Germanwings and the Lufthansa Passenger Airline are available until 2011, only. In 2012 both companies appear as a unit.

The Lufthansa Airline Group has been able to increase its number of passengers constantly from 55,5 million in 2009 to 65.4 million in 2011, which means an increase of almost 17 percent within a period of two years (figure 18).

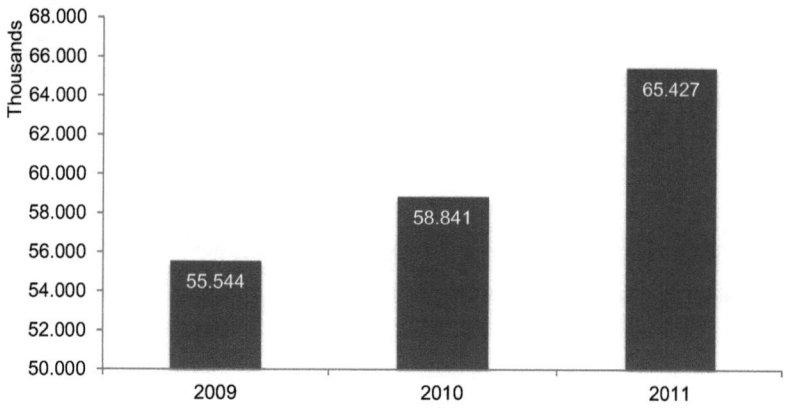

Figure 18:   Lufthansa Passenger Airline - Number of Passengers[147]

The same development can be observed for the Available Seat-Kilometers and the Revenue Passenger-Kilometers in figure 19. The demand rose according to the supply. Thus, the load factor stayed at a level of approximately 78 percent from 2009 to 2011.

---

[147] Own illustration on the basis of Lufthansa (2010a; 2011; 2012a).

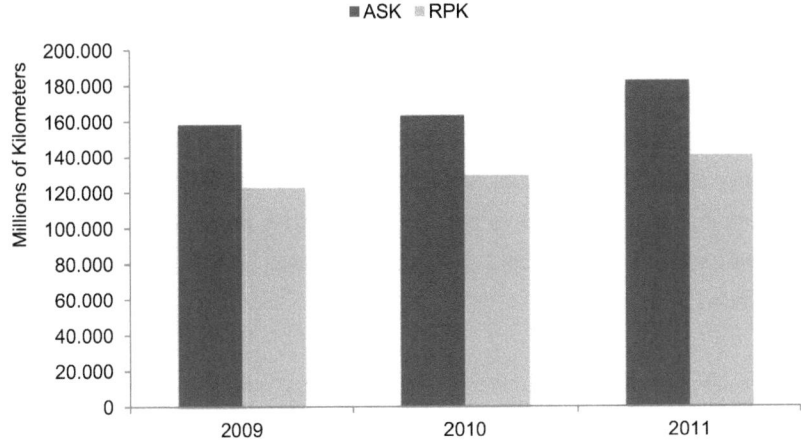

Figure 19: Lufthansa Passenger Airline - ASK and RPK in Millions[148]

The operating result, in comparison, features a decline since 2010 (figure 20). Whereas the airline was able to reach a profit of 382 million Euros that year, it had to suffer a negative result of 107 million Euros in 2011.

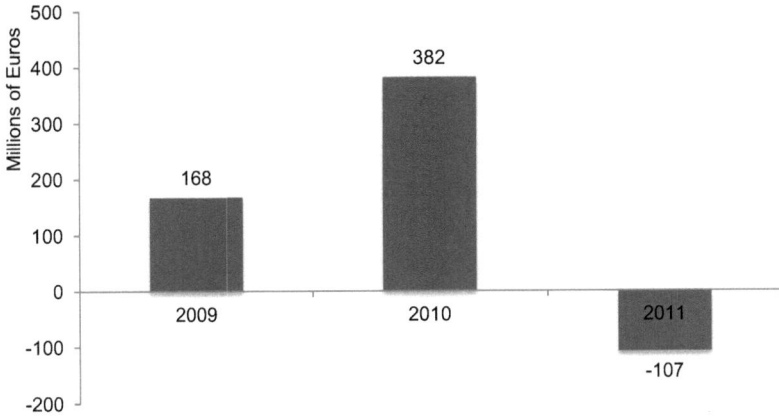

Figure 20: Lufthansa Passenger Airline - Operating Result[149]

---

[148] Own illustration on the basis of Lufthansa (2010a; 2011; 2012a).
[149] Own illustration on the basis of Lufthansa (2010a; 2011; 2012a)

6.4 The Subsidiary Germanwings and its Business Performance

The German airline Germanwings was founded in 2002 as a classical Low Cost Carrier (sub-chapter 3.2). Today it is operated by the Germanwings GmbH and the only shareholder is Deutsche Lufthansa. The Germanwings headquarter is located in Cologne, next to its base airport, which is the Cologne/Bonn airport. Further stations are located in Stuttgart, Berlin, Hamburg, Hanover and Dortmund.[150]

The airline has been able to increase its passenger numbers steadily from 2005 to 2008, but had to realize a decline in 2009 like the entire Lufthansa Passenger Airline Group and other airlines due to a decline in demand. One year later, in 2010, the numbers went up again and the airline reached a total of 829 million passengers. However, already in 2011 the number started to drop again (figure 21).

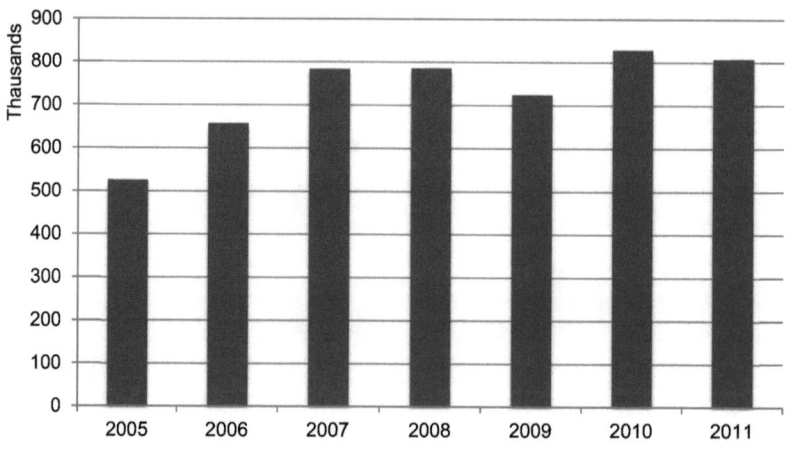

Figure 21: Germanwings - Number of Passengers from 2005 to 2011 in July[151]

---

[150] Germanwings (2013a).
[151] Own illustration on the basis of Germanwings (2006); Lufthansa (2011b).

The airline's load factor at the same time has always been above 80 percent (figure 22). Thus, it has been higher than the load factor of the total Lufthansa Passenger Airline Group. Nevertheless, is has never reached the load factors of the competing LCCs easyJet and Raynair, which usually reach a percentage above 85.[152]

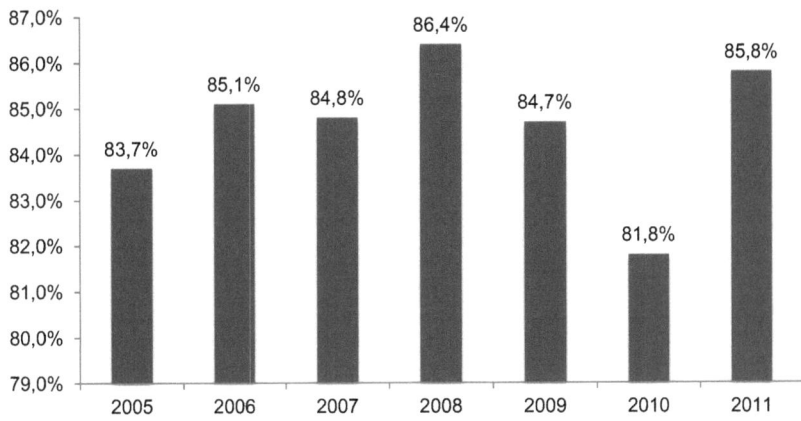

Figure 22: Germanwings - Load Factor from 2005 to 2011 in July[153]

The airline's operating income increased until 2009, but started dropping in 2010. Since that year the airline has not been able to realize a profit again. In 2011 the airline even experienced an operating loss of 52 million Euros (figure 23).

---

[152] easyJet (2013); Ryanair (2013).
[153] Own illustration on the basis of Germanwings (2006); Lufthansa (2011b).

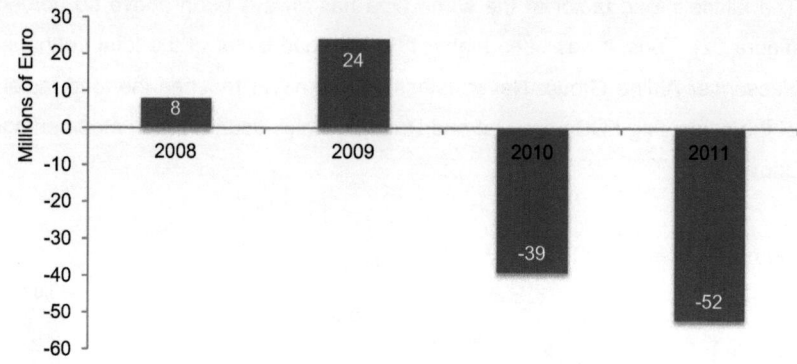

Figure 23: Germanwings - Operating Result[154]

More information about the business performance of Germanwings can be found in the appendix.

---

[154] Own illustration on the basis of Lufthansa (2009; 2010a; 2011; 2012a;).

7 The SWOT-Analysis applied for Lufthansa and Germanwings

According to the Lufthansa Chief Executive Officer Christoph Franz, the new business strategy implemented by the Lufthansa Passenger Airline and Germanwings is intended to combine the strengths of the two airlines. In order to achieve a better business performance, the new concept was created to realize lower operating costs than the Lufthansa Passenger Airline on its point-to-point flights within Europe, and to make the LCC Germanwings profitable.[155] For simplification reasons, this chapter uses abbreviations for the airlines, which are LH for the Lufthansa Passenger Airline and GW for Germanwings.

In order to appoint the strengths and weaknesses of both companies and the existing opportunities and threats of the macro-environment, the SWOT-analysis is approached in this sub-chapter. The foundation of the analysis is the previous chapters. The SWOT-analysis is used as it combines given statements in a clear manner and contrasts them.[156]

---

[155] Münck (2012).
[156] Asum and Stich (2009), p. 187-191.

Strengths	Weaknesses
• Strong brands (LH and GW) • Two business concepts (FSNC and LCC) within one company • Slots at coordinated airports through grandfather rights (LH) • Star Alliance membership (LH) • Entry to global network through alliance (GW) • Growing market share of Star Alliance • Existing route network • As part of the Lufthansa Group not dependent on external suppliers in every situation (catering) • Existing fleet (LH & GW) • Frequent flyer network through frequent flyer program (LH) • Existing suppliers and contracts	• Decreasing operating result (LH & GW) • High operating expenses (LH) • Stagnation in growth of the passenger number (GW) • Low load factor (GW)
Opportunities	**Threats**
• Expected market growth • German passenger airline market biggest one within Europe • Less difficult entry of foreign markets through Single Common Aviation Area • Decrease of competition through planned reduction of subsidies at regional airports • Decrease of competition from outside the European Union through the implementation of emission trading • Expected growth of GDP	• High number of competing airlines • Intense price competition • Low product differentiation among the competition • Increase in competition through Single Common Aviation Area • Higher costs through Aviation Tax Act • Higher costs through emission trading • Emerging airlines • Downturn in the demand for LCC • Growing expenses (fuel, fees, etc.) • Growing market share of competing alliance oneworld • Competition by long-haul buses and new rail connections within a distance of 800 to 1.000 kilometers

Table 4: SWOT-Portfolio Lufthansa Passenger Airline and Germanwings[157]

---

[157] Own illustration.

As shown in the table above both companies, LH as well as GW, have a number of strengths through which they might build up competitive advantages. These strengths comprise, for example, two well known strong brands, which the airlines have developed since they were founded. Whereas the brand Lufthansa mainly stands for quality[158], the brand Germanwings focuses on low fares.[159]

Additionally, both airlines have their own fleets of aircrafts as well as established route networks. Furthermore, LH is a member of the strategic alliance Star Alliance, which has been able to increase its share on the German market recently. Through this membership LH is able to offer flights to destinations which it does not approach itself.

On the other hand, both airlines had to experience weaknesses during the last years, such as decreasing operating results caused by high operating expenses, a stagnation of the passenger numbers and a weak load factor. Although the list of weaknesses is short compared to the number of strengths, their impact on the companies' performance is strong.

Nevertheless, there are opportunities which the German passenger market offers to LH and GW. These include, for example, the position of being one of the biggest airline markets within Europe according to the market value, with an expected further growth due to a growing Gross Domestic Product. Additionally, the entry of foreign markets has become easier through the creation of the Single Common Aviation Area.

Anyhow, there are also threats, which may influence the performance of LH and GW. A high number of competitors in addition to an intense price competition put a lot of pressure on both airlines. The growing market share of the strategic alliance oneworld and the emerging competition through substitute products intensify this force once more. Additionally, both airlines have to cope

---

[158] Hertle and Frühwald (2006).
[159] Lufthansa (2012c).

with higher costs through the Aviation Tax Act, emission trading and growing expenses in general, such as fuel prices and fees.

7.1 The New Germanwings – a Strategy Change

In 2013 the Lufthansa Passenger Airline and its subsidiary Germanwings implemented the restructured Low Cost Carrier Germanwings as the New Germanwings (NGW)[160]. Its head office is located in Cologne with further stations in Berlin, Dortmund, Dusseldorf, Hamburg, Hanover, and Stuttgart.

The reformed LCC is presented as a combination of the existing brands and products of the Lufthansa Passenger Airline's point-to-point-flights and the existing airline Germanwings. The new airline bundles the domestic and European point-to-point-flights by Lufthansa (except the flights from/to Frankfurt/Main and Munich) and the Germanwings route network. Nowadays, it serves more than 110 destinations within Europe.[161]

The business concept of the NGW includes three different products. These are BASIC, SMART and BEST (table 5), which are described more detailed below.

The BASIC product includes the core product only, which is the flight from destination A to destination B. By booking this product the traveler is able to join a frequent flyer program. The ticket is available for a low fare of 33 Euros, but can only be booked online on the airline's website. Extra services, such as luggage and catering, may be purchased additionally. Thus, the product BASIC is structured like the LCC product (sub-chapter 3.2). It is intended to attract people for whom the price matters most.

The SMART product can be seen as the extended version of the BASIC product. It does not only include the core product, but also some augmented services, such as a the reservation of a preferred seat, a snack and a drink

---

[160] The airline is still named Germanwings. However, in order to distinguish between the old and the new business concept, the following pages will use the name "New Germanwings" referring to the new business concept and the name "Germanwings" or rather the abbreviation GW referring to the old business concept.
[161] Lufthansa (2012c).

during the flight, and one piece of luggage up to 23 kg. This product contains the same features as the economy class of the classical FSNC usually contains. Ticket prices start at 53 Euros when booking it on the airline's website and at 68 Euros when booking it via a GDS (15 Euros GDS fee).

The product BEST contains even more services, which are, for example, the provision of a free middle seat, a reserved carry-on luggage tray, à-la-carte-catering, two pieces of luggage, etc. This product replaces the Lufthansa business class, which was offered by the Lufthansa Passenger Airline on the point-to-point-flights within Europe. This product is intended to attract business traveler. It is available for a price starting at 199 Euros if purchased on the airline's website and for 214 Euros if purchased via a GDS (15 Euros GDS fee).[162]

---

[162] Lufthansa (2012c); Jegminat (2013).

	BASIC	SMART	BEST
Product	• Flight from A to B • Frequent flyer program (Boomerang Club or Miles and More) • Extra services may be purchased additionally	• Flight from A to B • Frequent flyer program (Boomerang Club or Miles and More) • Seat reservation from row 4 on • Snacks and drinks • 1 piece of luggage of 23 kg	• Flight from A to B • Frequent flyer program double miles (Boomerang Club or Miles and More) • Seat reservation in row 1 to 3 • Free middle seat • Reserved carry-on luggage tray • À-la-carte-catering • 2 pieces of luggage of 23 kg • Security fast lane use • Lounge entrance • Change of flight option • Priority check-in • Priority boarding
Price	€ 33 and more	€ 53 and more via Website € 68 and more via GDS	€ 199 and more via Website € 214 and more via GDS
Place	Website	Website GDS	Website GDS
Promotion	New conjoint brand including new Corporate Design Facebook Twitter		

Table 5: The Business Concept of the NGW[163]

---

[163] Own illustration on the basis of Lufthansa (2012c); Germanwings (2013b; 2013c); Jegminat (2013); Miles & More (2013).

7.2 Evaluation of the New Strategy in Accordance to the SWOT-Analysis

The previous pages have demonstrated that the German passenger airline market offers some opportunities but also a number of threats for airlines. The Lufthansa Passenger Airline and Germanwings have tried to overcome their weaknesses by combining their strengths into a new business concept, which was described in sub-chapter 7.1. The following pages are going to evaluate whether this concept has a potential for success.

According to its characteristics, the NGW can be seen as a mixture of a FSNC and a LCC, although the airline calls itself a LCC. The airlines are trying to bundle their knowledge and experiences of the operating carrier types in order to create a business model that has not existed in this form before.

By offering products for low, medium and high fares, the NGW tries to attract all existing target groups, which are described in chapter 5. Since the numbers of private as well as of business travelers are expect to grow further, the airline may realize an increase in passenger numbers. On the other hand, the rivalry amongst competitors is high due to the possible market entry of new airlines through the creation of the Single Common Aviation Area and the emergence of new airlines (chapter 4). Additionally, the market share of LCC on the German passenger airline market has dropped since 2010, whereas FSNC have always held the biggest share with a potential for increase since 2010. Consequently, it can be said that the restructured airline was implemented as a business concept which is on the downgrade. Nevertheless, there are some features through which the airline may succeed against its competitors.

Contrary to other LCC, the NGW offers two frequent flyer programs in order to raise customer loyalty. These are the Boomerang program for frequent flyers using the NGW and Miles and More for frequent flyers using the NGW, LH and/or other airlines and partners.[164] Thereby, the airline might bind travelers. However, as it was stated in sub-chapter 4.4, costs for switching the airline are

---

[164] Germanwings (2013b); Miles & More (2013).

low. For private travelers, for example, the price matters most when booking a flight (sub-chapter 5.1). For this reason they are likely to switch the airline according to the lowest fare even though a frequent flyer program exists.

Both, LH and GW have developed independent strong brands as well as images during the past years. Whereas Lufthansa is a FSNC, which stands for high quality[165], GW is a LCC, which mainly promotes itself through low fares. By putting together these strengths, the NGW may create an image of a quality LCC. However, the combination might also result in a confusion and disappointment of travelers on the other hand. Business travelers, for example, who used to travel with LH, might stop using the carrier, as they might not be willing to travel with a LCC. Thus, combining both brands into a new one might provide some opportunities, but also some risks.

While the NGW is not planning on becoming a member of a strategic alliance, it may benefit from the membership of LH in the Star Alliance by offering connecting flights to the Star Alliance network. Thus, the NGW offers flights to destinations, which it does not approach itself. Furthermore, sub-chapter 4.1.2 illustrates the increasing market share of the Star Alliance on the German market, which might be caused by the joining of more airlines. Consequently, the route network will grow as well, which may lead to a higher number of destinations offered by the NGW as well.

A further advantage of the NGW is being a part of the Lufthansa Passenger Airline Group. The NGW, thus, is able to offer code sharing flights with Austrian Airlines and SWISS[166], which extends its route network yet again. Moreover, the Single Common Aviation Area might lead to a higher competition on the one hand; it might also lead to the extension of the route network throughout Europe on the other hand.

---

[165] Hertle and Frühwald (2006).
[166] Jegminat (2013).

Consequently, the NGW already offers flights to a wide range of international destinations through which it is able to realize a competitive advantage over other LCC and will be able to extend this network further.

Additionally, the NGW is a part of the Lufthansa Group, which also runs the business segments Maintenance Repair Overhaul, Catering, IT-Services and its own provider of training courses for airlines. Hence, the airline does not depend on external suppliers regarding technical, IT and catering matters. Furthermore, it is able to train its own crew and, thus, does not rely on externally trained personnel, which saves the airline costs.

The NGW may also benefit from existing contracts with suppliers, such as aircraft manufacturers, in the way of negotiating better conditions and realizing economies of scale.

As stated in sub-chapter 4.2.1 the allocation of slots at Germany's coordinated airports is subject to a number of different conditions, such as grandfather rights. Here, the NGW may use the advantages of LH, which owns some of these rights. In that way the airline can easily approach airports, which may be hard to access for other new entrants. However, fees at these airports are higher than at regional airports, such as Paderborn or Dortmund, which are mainly approached by LCC. Thus, these airlines might be able to offer their tickets for lower fares. The reduction of subsidies at regional airports might, however, stop this phenomenon and reduce the competition of the NGW as well.

By keeping the structure of a LCC, the airline is supposed to reduce its costs and to increase its operating income in return. These characteristics include, for example, the operation of a homogeneous fleet (Airbus A319 as the only aircraft), simple processes and lower labor costs. Although the NGW is able to reduce the operating costs itself, there are some increasing costs, which it cannot influence. These, for example, include costs caused by the Aviation Tax Act (sub-chapter 4.2.4), by the implementation of emission trading, as well as by rising fuel prices.

Finally, it can be said that LH and GW identified their weaknesses and try to overcome them by combining their strengths. They put together their strong brands, their existing fleets and route networks and created the NGW with the cost structure of an LCC. Thus, they might be able to realize some cost advantages. However, the concept will only overcome the LH's and GW's deficits if it asserts itself on the passenger market. Taking into consideration the threats of the existing intense price competition and of low product differentiation, it is doubtful whether the concept, which focuses on the price, will become competitive.

8   Conclusion and Recommendations

The purpose of this paper has been to give an overview of the recent situation of the German passenger airline market and to prove the potential for economic success for the new business concept applied by the German Lufthansa Passenger Airline and Germanwings. Therefore, this paper was divided into three main areas.

First, the theoretical part discussed the marketing mixes of Full Service Network Carriers, Low Cost Carriers, Regional Carriers and Leisure Carriers, pointing out their competitive advantages and disadvantages. Consequently, it was said that the classical business concepts rarely exist anymore, since airlines try to combine features of all concepts.

The second area comprised the analysis part. Porter's five forces were applied in order to evaluate the rivalry among the direct competition, the threat of new entry and substitute products, as well as the bargaining power of the buyers and the suppliers. It was indicated that the rivalry among existing competitors is very strong due to a number of about 150 different carriers, whereas two-thirds of all flights are operated by FSNC and a quarter by LCC.
The threat of new entry was evaluated as low due to a number of barriers, such as the Aviation Tax Act, which imposes fees on flight tickets, and the allocation process of slots at German airports. Furthermore, it was shown that the emergence of substitute products and services increases the existing competition on a distance of 800 to 1.000 kilometers. However, this has been relatively weak, since these substitute products lose their competitiveness on distances farther than 1.000 kilometers.
The list of suppliers is short, which increases their bargaining power to a high level and lowers the bargaining power of the airlines in return. In comparison, the bargaining power of the buyer is balanced due to a high number of buyers on the one side and low costs of switching the airline on the other side. According to these findings it was said, that the German passenger market is

not attractive for airlines to operate in. Yet, experts forecast a growing demand for flights.

Subsequently, the paper focused on the LH and its subsidiary GW. In order to evaluate their conjunct implemented business concept of the NGW, both airlines' strengths and weaknesses were highlighted. It appeared that two strong brands, two existing fleets and route networks, as well as an access to the Star Alliance route network on the one hand, were facing decreasing operating results of both airlines due to high operating expenses and a relatively low load factor on the other hand.

Adherently, the findings were contrasted using the SOWT-analysis and the implemented business concept of the NGW was evaluated. Consequently, it seemed that by combining some of the LH's and the GW's strengths, such as the existing brand images and the existing route networks, the NGW tries to eliminate the previous concepts' weaknesses. However, it appears that some threats, which have been on the market, were not taken into consideration while creating a new business concept. These include an already existing price competition, the downgrade of the LCC business concept and a low product differentiation. Since these threats may have a strong influence on the NGW'S performance, it is doubtful whether the NGW will become competitive.

Although the LCC market share has dropped slightly during the last years, its portion is still high. The same appears to be true for the FSNC, which was even able to increase its market share during the last years. Now that different LCC are adding services to their basic products and FSNC are excluding them, it may be time to return to the roots in order to stay competitive.

Therefore, it might have been better to implement the NGW as a classic LCC with simple structures and some added features. This could have been the product BASIC (table 5) along with a distribution via a GDS. In that way, LH and GW might have been able to reduce their operating expenses and to increase their operating income in return. Moreover characteristics, such as an extensive route network through the Star Alliance membership of LH and a frequent flyer program on the one hand and low fares on the other hand would have made the NGW competitive.

Appendix

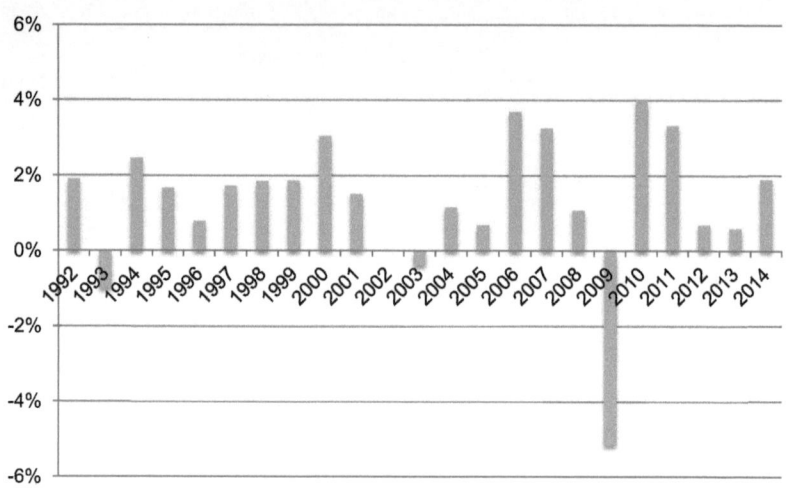

Figure 24: Development of the GDP in Germany and Forecast [167]

	Market Value in $ Billion	Percentage
Germany	25,6	14,0
United Kingdom	24,2	13,2
Spain	20,4	11,1
Italy	18,3	10,0
France	12,6	6,9
Rest of Europe	82,1	44,8
Total	183,2	100

Table 6: Geographical Segmentation of the European Passenger Airline Market in 2011 [168]

	2009	2010	2011

---

[167] Own illustration on the basis of Statistisches Bundesamt (2013); Statista (2013).
[168] Own illustration on the basis of MarketLine (2012).

Full Service Network Carriers	63,3 %	61,5 %	64,3 %
Low Cost Carriers	31,0 %	30,3 %	28,5 %
Leisure Carriers	3,1 %	6,1 %	5,5 %
Regional Carriers	2,5 %	2,2 %	1,7 %

Table 7: Market Share of Existing Business Concepts[169]

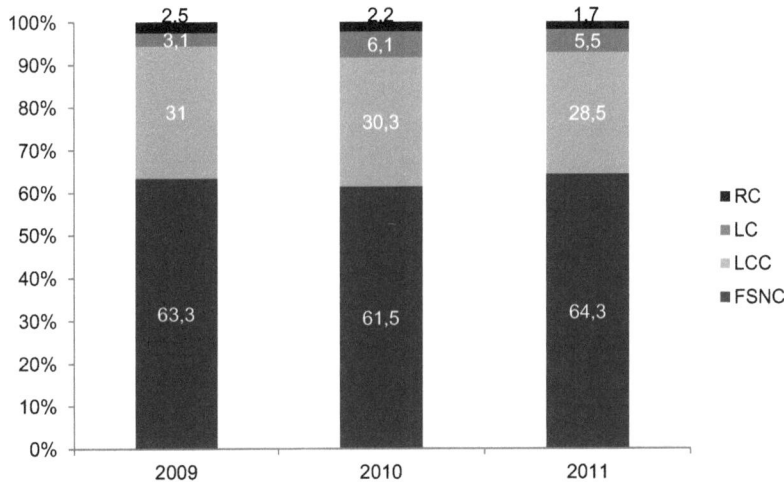

Figure 25: Market Share of Existing Business Concepts[170]

---

[169] Own illustration on the basis of Deutsches Zentrum für Luft- und Raumfahrt e.V. (2011; 2012).
[170] Own illustration on the basis of Deutsches Zentrum für Luft- und Raumfahrt e.V. (2011; 2012).

	2009	2010	2011
Star Alliance	78,5 %	80,9 %	81,2 %
Independent	8,8 %	7,4 %	7,1 %
SkyTeam	8,4 %	7,4 %	4,1 %
oneworld	4,3 %	4,2 %	7,5 %

Table 8: FSNC Supply According to Alliance Membership[171]

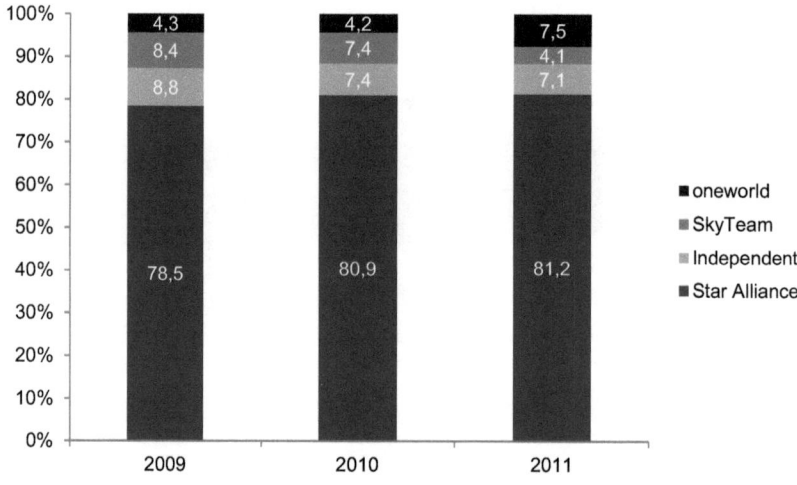

Figure 26: FSNC Supply According to Alliance Membership[172]

---

[171] Own illustration on the basis of Deutsches Zentrum für Luft- und Raumfahrt e.V. (2011; 2012).
[172] Own illustration on the basis of Deutsches Zentrum für Luft- und Raumfahrt e.V. (2011; 2012).

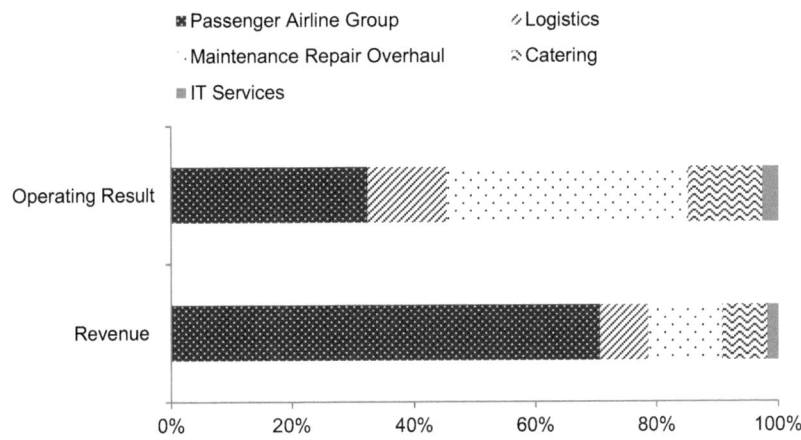

Figure 27: The Lufthansa Group Business Segments - Share of the Operating Result and the Revenue in 2012[173]

---

[173] Own illustration on the basis of Lufthansa (2013a).

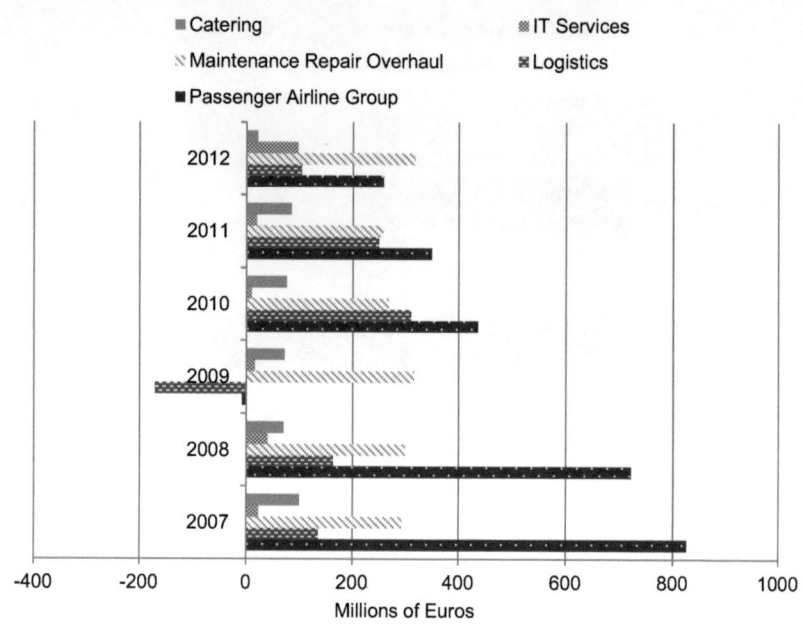

Figure 28: Operating Result of the Lufthansa Group's Business Segments[174]

Year	Number of Passengers (Thousands)	ASK (Millions)	RPK (Millions)	Load Factor
2012	103.051	259.861	204.775	78,8%
2011	100.605	258.263	200.376	77,6%
2010	91.157	235.837	187.000	79,3%
2009	76.543	206.269	160.647	77,9%
2008	70.543	195.431	154.156	78,9%
2007	62.878	169.108	135.011	79,8%

Table 9: Lufthansa Passenger Airline Group - Business Performance[175]

---

[174] Own illustration on the basis of Lufthansa (2008; 2009; 2010a; 2011; 2012a; 2013a).
[175] Own illustration on the basis of Lufthansa (2008; 2009; 2010a; 2011; 2012a; 2013a).

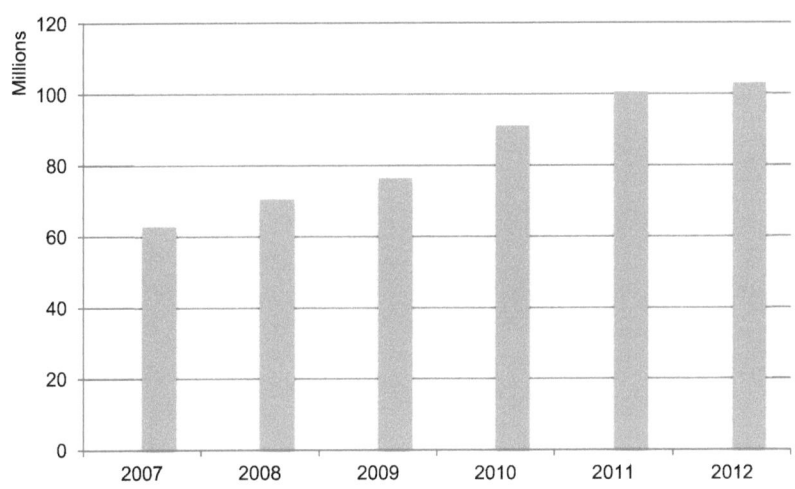

Figure 29: Lufthansa Passenger Airline Group - Number of Passengers[176]

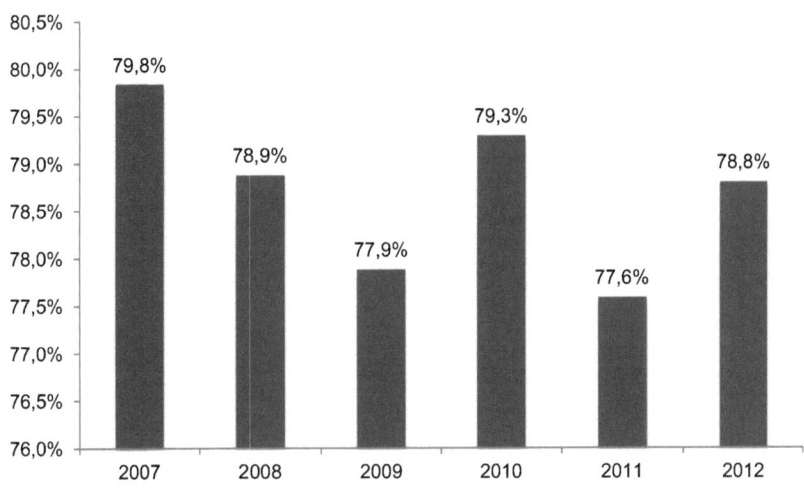

Figure 30: Lufthansa Passenger Airline Group - Load Factor from 2007 to 2012[177]

---

[176] Own illustration on the basis of Lufthansa (2008; 2009; 2010a; 2011; 2012a; 2013a).
[177] Own illustration on the basis of Lufthansa (2008; 2009; 2010a; 2011; 2012a; 2013a).

	2007	2008	2009	2010	2011	2012
Materials and services	9.270	11.510	10.904	13.250	14.542	15.749
Fuel	3.378	4.806	3.381	4.793	5.769	6.870
Fees	2.886	3.234	3.552	4.335	4.731	4.913
MRO services	1.341	1.495	1.790	1.869	1.909	1.782
Operating lease	199	281	338	247	136	113
Staff	2.959	3.025	3.330	3.829	3.874	3.945
Depreciation and amortization	822	929	1.032	1.261	1.350	1.415
Other operating expenses	2.991	2.812	2.805	3.310	3.280	3.119
Agency commissions	512	491	356	433	359	357
External staff	71	479	499	539	577	609
Rental and maintenance	335	179	209	0	0	0
Total operating expenses	**16.042**	**18.276**	**18.071**	**21.650**	**23.046**	**24.228**

Table 10: Lufthansa Passenger Airline Group - Operating Expenses in m€[178]

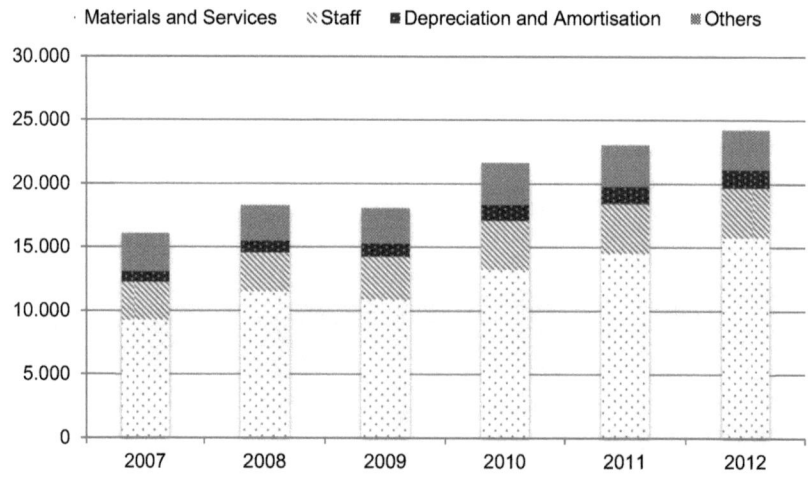

Figure 31: Lufthansa Passenger Airline Group - Operating Expenses in m€[179]

---

[178] Own illustration on the basis of Lufthansa (2008; 2009; 2010a; 2011; 2012a; 2013a).
[179] Own illustration on the basis of Lufthansa (2008; 2009; 2010a; 2011; 2012a; 2013a).

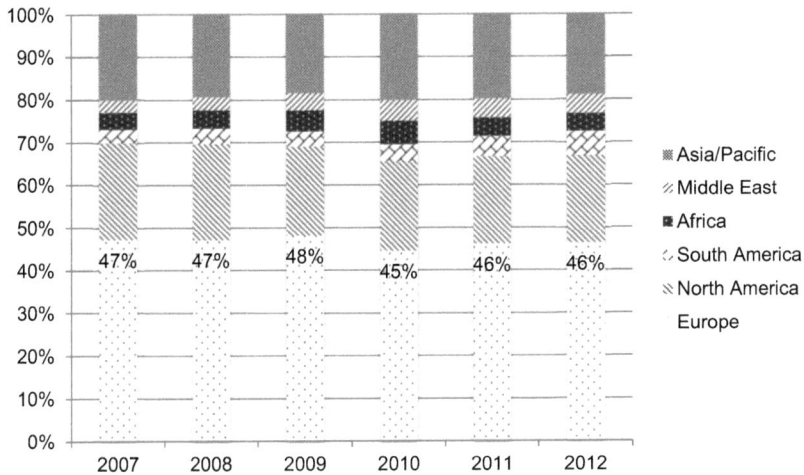

Figure 32: Lufthansa Passenger Airline Group - Regional Market Share[180]

Year	Number of Passengers (Thousands)	Load Factor	Load Factor
2011	7.522	85,8 %	85,8 %
2010	7.730	81,8 %	81,8 %
2009	7.166	84,7 %	84,7 %
2008	7.623	86,4 %	86,4 %

Table 11: Germanwings - Business Performance[181]

---

[180] Own illustration on the basis of Lufthansa (2008; 2009; 2010a; 2011; 2012a; 2013a).
[181] Own illustration on the basis of Lufthansa (2009; 2010a; 2011; 2012a).

List of References

Books, Book Sections and Videos

Asum, Heiko; Stich, Volker (2009): Die Besten Strategietools in Der Praxis: Welche Werkzeuge Brauche Ich Wann? Wie Wende Ich Sie an? Wo Liegen Die Grenzen? Munich: Hanser.

Bruhn, Manfred (2012): Marketing: Grundlagen Für Studium Und Praxis. Wiesbaden: Gabler.

Conrady, Roland; Fichert, Frank; Sterzenbach, Rüdiger (2013) Luftverkehr: betriebswirtschaftliches Lehr- und Handbuch. Munich: Oldenbourg.

Doganis, Rigas (2010): Flying Off Course: Airline Economics and Marketing. 4th ed. London; New York: Routledge.

Freyer, Walter (2009): Tourismus-Marketing: marktorientiertes Management im Mikro- und Makrobereich der Tourismuswirtschaft. Munich: Oldenbourg.

Fundamentals of Pricing and Revenue Management (2009) *In* The Global Airline Industry Pp. 73–111. Chichester, West Sussex, U.K.: Wiley.

Groß, Sven (2007): Handbook of Low Cost Airlines: Strategies, Business Processes and Market Environment. Berlin: Erich Schmidt.

Groß, Sven; Schröder, Alexander (2005):Low Cost Airlines in Europa: eine marktorientierte Betrachtung von Billigfliegern. Dresden: FIT-Forschungsinst. für Tourismus.

Maurer, Peter (2006): Luftverkehrsmanagement: Basiswissen. Munich [u.a.]: Oldenbourg.

Meffert, Heribert (2012): Marketing: Grundlagen marktorientierter Unternehmensführung. Gabler.

Michael Porter- On Five Forces Model (2011). http://www.youtube.com/watch?v=mYF2_FBCvXw, accessed August 27, 2013.

Pompl, Wilhelm (2007): Luftverkehr eine ökonomische und politische Einführung. Berlin: Springer.

Ruperti, Florian (2012): Marketing von Low-Cost-Airlines: Analyse der Gestaltungsoptionen des Marketingprogrammes von Low-Cost-Airlines im deutschen Markt im Hinblick auf den Kundennutzen mit dem Ziel einer nachhaltigen Verbesserung der Wettbewerbsposition. Munich: Hampp.

Schulz, Axel (2010): Globale Distributionssysteme. *In* Informationsmanagement Im Tourismus: E-Tourismus: Prozesse und Systeme Pp. 264–289. Munich: Oldenbourg.

Shaw, Stephen (2011): Airline Marketing and Management. 7th ed. Burlington, VT: Ashgate.

Tourismus Und Verkehr: Grundlagen, Marktanalyse Und Strategien von Verkehrsunternehmen (2011). Munich: Oldenbourg.

Magazine, Newspaper and Journal Article

Döring, Tobias (2013): Kein Steuergeld mehr für Provinzflughäfen. Handelsblatt, July 3. http://www.handelsblatt.com/unternehmen/handel-dienstleister/eu-will-subventionen-kappen-kein-steuergeld-mehr-fuer-provinzflughaefen/v_detail_tab_print/8443776.html, accessed August 13, 2013.

During, Rainer W. (2012): Lufthansa und AirBerlin kämpfen gemeinsam. Der Tagesspiegel, March 3. http://www.tagesspiegel.de/wirtschaft/luftverkehrssteuer-lufthansa-und-air-berlin-kaempfen-gemeinsam/v_print/6276592.html?p=, accessed August 13, 2013.

Graue, Oliver (2013): Sie können auch "Basic", FVW Magazin, July 5: 44–45.

Hegmann, Gerhard (2013): Chinas teure Luftnummer. Welt Online, September 8. http://www.welt.de/print/welt_kompakt/print_wirtschaft/article118841754/Chinas-teure-Luftnummer.html, accessed August 28, 2013.

Jegminat, Georg (2013): So tickt die neue Germanwings. FVW Magazin, April 26: 42–43.

Kamann, Matthias (2013): Die Fernbusse in Deutschland boomen. Welt Online, August 25. http://www.welt.de/politik/deutschland/article119351254/Die-Fernbusse-in-Deutschland-boomen.html, accessed August 27, 2013.

Kotowski, Timo (2012): Emissionshandelssystem EU setzt Klimaschutzabgabe auf Flüge aus. FAZ.NET, December 11. http://www.faz.net/aktuell/wirtschaft/wirtschaftspolitik/emissionshandelssystem-eu-setzt-klimaschutzabgabe-auf-fluege-aus-11958195.html#Drucken, accessed August 14, 2013.

Loppow, Bernd (1997): Die Liberalisierung Des Europäischen Luftmarktes: Was Sind die Folgen? Totaler Wettbewerb in Europa: Nach Ostern machen neue Fluggesellschaften auch der Lufthansa Konkurrenz. ZEIT Online, March 28. http://www.zeit.de/1997/14/opensky.txt.19970328.xml, accessed August 9, 2013.

Münck, Rita (2012): Der Brombeer-gelbe Weg. FVW Magazin, October 19: 30–31.

Opfermann, Matthias (2012): Online im Flugzeug: Bei diesen Airlines kann man im Flieger surfen - Fliegen. FOCUS Online. http://www.focus.de/reisen/flug/surfen-im-flieger-airlines-im-wettstreit-um-internetloesungen-_aid_750764.html, accessed July 23, 2013.

Spaeth, Andreas (2012): Regionalflughäfen in der Provinz beginnt's. FAZ.NET, October 28. http://www.faz.net/aktuell/reise/nah/regionalflughaefen-in-der-provinz-beginnt-s-11941243.html, accessed August 29, 2013.

Reports, Documents, Presentations, Emails

AIRBUS (2012): Navigating the Future. Global Market Forecast 2012-2031. http://www.airbus.com/company/market/forecast/?eID=dam_frontend_push&docID=27599, accessed August 23, 2013.

Arbeitsgemeinschaft Deutscher Verkehrsflughäfen (2002): ADV-Monatsstatistik Dezember 2001. Stuttgart. http://www.adv.aero/fileadmin/pdf/statistiken/2001/Statistik_2001_dez_adv.pdf, accessed August 21, 2013.

Arbeitsgemeinschaft Deutscher Verkehrsflughäfen (2003): ADV-Monatsstatistik Dezember 2002. Stuttgart. http://www.adv.aero/fileadmin/pdf/statistiken/2002/Statistik_2002_dez_adv.pdf, accessed August 21, 2013.

Arbeitsgemeinschaft Deutscher Verkehrsflughäfen (2004): ADV-Monatsstatistik Dezember 2003. Stuttgart. http://www.adv.aero/fileadmin/pdf/statistiken/2003/Statistik_2003_dez.pdf, accessed August 21, 2013.

Arbeitsgemeinschaft Deutscher Verkehrsflughäfen (2005): ADV-Monatsstatistik Dezember 2004. Berlin. http://www.adv.aero/fileadmin/pdf/statistiken/2004/Statistik_2004_dez.pdf, accessed August 21, 2013.

Arbeitsgemeinschaft Deutscher Verkehrsflughäfen (2006): ADV-Monatsstatistik Dezember 2005. Berlin. http://www.adv.aero/fileadmin/pdf/statistiken/2005/Statistik_2005_dezember.pdf, accessed August 21, 2013.

Arbeitsgemeinschaft Deutscher Verkehrsflughäfen (2007): ADV-Monatsstatistik Dezember 2006. Berlin. http://www.adv.aero/fileadmin/pdf/statistiken/2006/Statistik_Dezember_2006.pdf, accessed August 21, 2013.

Arbeitsgemeinschaft Deutscher Verkehrsflughäfen (2008): ADV-Monatsstatistik Dezember 2007. Berlin. http://www.adv.aero/fileadmin/pdf/statistiken/2007/Statistik_Dezember_2007.pdf, accessed August 21, 2013.

Arbeitsgemeinschaft Deutscher Verkehrsflughäfen (2009): ADV-Monatsstatistik Dezember 2008. Berlin. http://www.adv.aero/fileadmin/pdf/statistiken/2008/Statistik_Dezember_2008.pdf, accessed August 21, 2013.

Arbeitsgemeinschaft Deutscher Verkehrsflughäfen (2010): ADV-Monatsstatistik Dezember 2009. Berlin. http://www.adv.aero/fileadmin/pdf/statistiken/2009/Dezember_2009.pdf, accessed August 21, 2013.

Arbeitsgemeinschaft Deutscher Verkehrsflughäfen (2011): ADV-Monatsstatistik Dezember 2010. Berlin. http://www.adv.aero/fileadmin/pdf/statistiken/2010/ADV_Monatsstatistik_Dez_2010_fin al.pdf, accessed August 21, 2013.

Arbeitsgemeinschaft Deutscher Verkehrsflughäfen (2012): ADV-Monatsstatistik Dezember 2011. Berlin. http://www.adv.aero/fileadmin/pdf/statistiken/2011/ADV-Monatsstatistik_Dezember_2011.pdf, accessed August 21, 2013.

Arbeitsgemeinschaft Deutscher Verkehrsflughäfen (2013a): ADV-Monatsstatistik Dezember 2012. Berlin. http://www.adv.aero/fileadmin/pdf/statistiken/2012/12.2012_ADV-Monatsstatistik.pdf, accessed August 21, 2013.

Arbeitsgemeinschaft Deutscher Verkehrsflughäfen (2013b): ADV-Monatsstatistik Juni 2013. Berlin. http://www.adv.aero/fileadmin/pdf/statistiken/2013/06.2013_ADV-Monatsstatistik.pdf, accessed August 21, 2013.

Berster, Peter (2012): Entwicklung der globalen Luftfahrtmärkte - Passagiere, Transportwege, Zukunft. Paper presented at the Investment Conference, Frankfurt, May 10. http://www.scope.de/scope/download/scope_circles/unterlagen/Transport_20120510/0 4_CTransport_DLR_DrBerster.pdf, accessed September 22, 2013.

BUND für Umwelt und Naturschutz Deutschland e.V. (2012): Wirkungen Der Luftverkehrssteuer in Deutschland: Eine Erste Einschätzung. Berlin. http://www.bund.net/fileadmin/bundnet/pdfs/verkehr/luftverkehr/20120301_verkehr_luft verkehrssteuer_einschaetzung.pdf, accessed August 13, 2013.

Bundesministerium für Umwelt, Naturschutz und Reaktorsicherheit (2013): Emissionshandel für Klimaschutz und Energiewende. http://www.bmu.de/fileadmin/Daten_BMU/Download_PDF/Emissionshandel/faltblatt_e missionshandel_bf.pdf, accessed August 14, 2013.

Centrum für Europäische Politik (2012): Slotzuweisung auf EU-Flughäfen. http://www.cep.eu/fileadmin/user_upload/Kurzanalysen/Slotzuweisung/KA_Slotzuweisu ng_auf_EU-Flughaefen.pdf, accessed August 12, 2013.

Deutsche Flugsicherung (2013): Luftverkehr in Deutschland - Mobilitätsbericht 2012. http://www.dfs.de/dfs_homepage/de/Presse/Publikationen/Mobilitaetsbericht_2012_We b.pdf, accessed August 18, 2013.

Deutscher Reiseverband e.V. (2013): Studie Business Travel 2013. Berlin. http://www.chefsache-businesstravel.de/fileadmin/user_upload/docs/DRV_130306_Studie_Business_Travel_ 2013_Kurz.pdf.

Deutsches Zentrum für Luft- und Raumfahrt e.V. (2011): Luftverkehrsbericht 2010 - Daten und Kommentierungen des deutschen und weltweiten Luftverkehrs. Cologne: Deutsches Zentrum für Luft- und Raumfahrt e.V. in der Helmholtz-Gemeinschaft. http://www.dlr.de/fw/Portaldata/42/Resources/images/luftverkehrsbericht2009/Luftverke hrsbericht2010.pdf, accessed September 27, 2013.

Deutsches Zentrum für Luft- und Raumfahrt e.V. (2012): Luftverkehrsbericht 2011 - Daten Und Kommentierungen des deutschen und weltweiten Luftverkehrs. Cologne: Deutsches Zentrum für Luft- und Raumfahrt e.V. in der Helmholtz-Gemeinschaft. http://www.dlr.de/fw/Portaldata/42/Resources/dokumente/pdf/LVB2011.pdf, accessed September 22, 2013.

Deutsches Zentrum für Luft- und Raumfahrt e.V. (2013): Low Cost Monitor 1/2013 - An Analysis Performed by DLR. Cologne. http://www.dlr.de/fw/Portaldata/42/Resources/dokumente/aktuelles/LCC_Monitor_I_201 3_en.pdf, accessed September 22, 2013.

Deutsches Zentrum für Luft- und Raumfahrt e.V.; Arbeitsgemeinschaft Deutscher Verkehrsflughäfen (2007): Low Cost Monitor 1/2007 - Eine Gemeinsame Untersuchung von DLR Und ADV. Cologne. http://www.dlr.de/fw/Portaldata/42/Resources/dokumente/aktuelles/Low_Cost_Monitor_ I_2007.pdf, accessed September 22, 2013.

Deutsches Zentrum für Luft- und Raumfahrt e.V.; Arbeitsgemeinschaft Deutscher Verkehrsflughäfen (2010): Low Cost Monitor 1/2010 - Eine Gemeinsame Untersuchung von DLR Und ADV. Cologne. http://www.dlr.de/fw/Portaldata/42/Resources/dokumente/aktuelles/Low_Cost_Monitor_ I_2010_final.pdf, accessed September 22, 2013.

Euopean Parliament; European Council (2008): Regulation (EC) No 1008/2008 of the European Parliament and of the Council of 24 September 2008 on Common Rules for the Operation of Air Services in the Community. http://eur-lex.europa.eu/LexUriServ/LexUriServ.do?uri=OJ:L:2008:293:0003:0020:en:PDF, accessed August 9, 2013.

European Communities (1993): Council Regulation (EEC) No 95/93 of 18 January 1993 on Common Rules for the Allocation of Slots at Community Airports. http://www.fhkd.org/images/pdf/VO-int.pdf, accessed August 8, 2013.

Flughafen Köln Bonn GmbH (n.a.): Fluggastbefragung Flughafen Köln Bonn.

Germanwings (2013a): Facts and Figures. http://www.germanwings.com/downloads/Germanwings_Facts_and_Figures.pdf, accessed October 8, 2013.

Hertle, Thomas; Frühwald, Sabine (2006): Die Marke Lufthansa. Nuremberg: Gesellschaft für Konsumforschung. http://www.gfk.com/imperia/md/content/gfkmarktforschung/maerkte/die_marke_lufthansa.pdf, accessed October 18, 2013.

Initiative Airport Media; Gesellschaft für Konsumforschung (2011): Airport Private Traveller Study. Reiseverhalten, Einstellungen und Werte der Privatreisenden am Airport. http://www.airport-media-muc.de/media/download/bereiche/mediacenter/extras/deutsch/Airport_Private_Traveller_Study.pdf, accessed August 23, 2013.

Lufthansa (2008): Geschäftsbericht 2007. Cologne. http://investor-relations.lufthansagroup.com/fileadmin/downloads/de/finanzberichte/geschaeftsberichte/LH-GB-2007-d.pdf, accessed October 5, 2013.

Lufthansa (2009): Geschäftsbericht 2008. Cologne. http://investor-relations.lufthansagroup.com/fileadmin/downloads/de/finanzberichte/geschaeftsberichte/LH-GB-2008-d.pdf, accessed October 5, 2013.

Lufthansa (2010a): Geschäftsbericht 2009. Cologne. http://investor-relations.lufthansagroup.com/fileadmin/downloads/de/finanzberichte/geschaeftsberichte/LH-GB-2009-d.pdf, accessed October 5, 2013.

Lufthansa (2010b): Nationale Luftverkehrssteuer verzerrt den Wettbewerb. http://www.lufthansagroup.com/fileadmin/downloads/de/politikbrief/07_2010/LH-Politikbrief-Juli-2010-Luftverkehrsabgabe.pdf, accessed August 13, 2013.

Lufthansa (2011a): Geschäftsbericht 2010. Cologne. http://investor-relations.lufthansagroup.com/fileadmin/downloads/de/finanzberichte/geschaeftsberichte/LH-GB-2010-d.pdf, accessed October 5, 2013.

Lufthansa (2011b): Investor Info July 2011. 07/11. Frankfurt. http://investor-relations.lufthansagroup.com/fileadmin/downloads/de/finanzberichte/verkehrszahlen/lufthansa/2011/LH-Investor-Info-2011-07-d.pdf, accessed October 21, 2013.

Lufthansa (2012a): Geschäftsbericht 2011. Cologne. http://investor-relations.lufthansagroup.com/fileadmin/downloads/de/finanzberichte/geschaeftsberichte/LH-GB-2011-d.pdf, accessed October 5, 2013.

Lufthansa (2012b): Kleinstflughäfen: Fragwürdigen Subventionswettlauf beenden. Politikbrief. http://www.lufthansagroup.com/fileadmin/downloads/de/politikbrief/10_2011/LH-Politikbrief-Oktober-2011-Kleinstflug%C3%A4fen.pdf, accessed August 10, 2013.

Lufthansa (2012c): Die Neue Germanwings - Check-in to the Smart Class. Cologne, December 6. http://www.lufthansagroup.com/fileadmin/themen/de/germanwings/121130_PK_Presentation_d.pdf, accessed October 16, 2013.

Lufthansa(2013a): Geschäftsbericht 2012. Cologne. http://investor-relations.lufthansagroup.com/fileadmin/downloads/de/finanzberichte/geschaeftsberichte/LH-GB-2012-d.pdf, accessed October 5, 2013.

Luftverkehrsteuergesetz (2010). Bundesministerium der Justiz. http://www.gesetze-im-internet.de/luftvstg/index.html, accessed August 13, 2013.

MarketLine (2012): Airlines in Europe. London.

Statistisches Bundesamt (2005): Publikation - STATmagazin - Verkehr - STATmagazin - Thema Flugzeug Oder Bahn - Womit geht es auf Reisen? https://www.destatis.de/DE/Publikationen/STATmagazin/Verkehr/2008_02/Flug_Eisenbahnverkehr.html, accessed August 17, 2013.

Statistisches Bundesamt (2013): Volkswirtschaftliche Gesamtrechnungen-Bruttoinlandsprodukt, Bruttonationaleinkommen, Volkseinkommen-Lange Reihen ab 1950. Wiesbaden. https://www.destatis.de/DE/ZahlenFakten/GesamtwirtschaftUmwelt/VGR/Inlandsprodukt/Inlandsprodukt.html, accessed September 30, 2013.

Statistisches Bundesamt (2013a): Wirtschaftsbereiche - Personenverkehr - Personenbeförderung. https://www.destatis.de/DE/ZahlenFakten/Wirtschaftsbereiche/TransportVerkehr/Personenverkehr/Tabellen/Flugpassagiere.html, accessed August 17, 2013.

Streule, Josef (2012): Subventionen für Regionalflughäfen: Billigflieger Profitieren. http://www.daserste.de/information/wirtschaft-boerse/plusminus/sendung/br/2012/28112012-regionalflughaefen100.html, accessed August 10, 2013.

Verband Deutsches Reisemanagement e.V. (2009): VDR Business Travel Report Germany 2009. 11. Frankfurt. http://www.vdr-service.de/fileadmin/fachthemen/geschaeftsreiseanalyse/vdr_btr2009.pdf, accessed August 23, 2013.

Verband Deutsches Reisemanagement e.V. (2012a): VCD Bahntest 2012/2013, Bahn-Flug-Kostencheck. Berlin. http://www.vcd.org/fileadmin/user_upload/redakteure_2010/projekte/vcdbahntest/VCD_Bahntest_2012.pdf, accessed August 16, 2013.

Verband Deutsches Reisemanagement e.V. (2012b): VCD Stellungnahme Luftverkehrssteuer. http://www.bundestag.de/bundestag/ausschuesse17/a07/anhoerungen/2012/110/Stellungnahmen/10-VCD.pdf, accessed August 13, 2013.

Verband Deutsches Reisemanagement e.V. (2013): VDR Business Travel Report Germany 2013. 11. Frankfurt. http://www.vdr-service.de/fileadmin/fachthemen/geschaeftsreiseanalyse/vdr_btr2013.PDF, accessed August 23, 2013.

Verkehrsclub Deutschland e.V. (2009): VCD Bahntest 2009, Die Mobilitätsbedürfnisse von Fahrgästen und potenziellen Fahrgästen der Bahn. Berlin. http://www.vcd.org/fileadmin/user_upload/redakteure_2010/projekte/vcdbahntest/vcd_bahntest2009_hintergrund.pdf, accessed August 20, 2013.

Websites

AirBerlin (2013): AirBerlin Group - Strategie und Geschäftsmodell. http://www.airberlingroup.com/de/ueber-airberlin/strategie-und-geschaeftsmodel, accessed September 25, 2013.

AIRBUS (2013): Dimensions & Key Data | Airbus, a Leading Aircraft Manufacturer. http://www.airbus.com/aircraftfamilies/passengeraircraft/a380family/specifications/, accessed August 28, 2013.

Airport Coordination Germany (2013a): Fluko Airport Coordination Germany. http://www.fhkd.org/index.php/facilitated, accessed August 8, 2013.

Airport Coordination Germany (2013b): Fluko Airport Coordination Germany. http://www.fhkd.org/index.php/coordinated, accessed August 8, 2013.

American Marketing Association (2007): Definition of Marketing. http://www.marketingpower.com/AboutAMA/Pages/DefinitionofMarketing.aspx, accessed July 15, 2013.

Boeing (1995): 747 Family. http://www.boeing.com/boeing/commercial/-747family/pf/pf_seating_charts.page, accessed August 28, 2013.

Bundesministerium für Verkehr, Bau und Stadtentwicklung (n.a.): Europäische Eisenbahnpolitik. http://www.bmvbs.de/DE/VerkehrUndMobilitaet/Verkehrstraeger/Schiene/Europaeische Eisenbahnpolitik/europaeische-eisenbahnpolitik_node.html;jsessionid=996B9C3F709B79956BB9995206972679, accessed August 27, 2013.

Deutsche Presse Agentur (2012): Fernbus-Liberalisierung ab 2013: Grünes Licht für die Bahn-Konkurrenz. Stern.de. http://www.stern.de/reise/deutschland/fernbus-liberalisierung-ab-2013-gruenes-licht-fuer-die-bahn-konkurrenz-1894877.html, accessed August 27, 2013.

easyJet (2013): Traffic Statistics. http://corporate.easyjet.com/investors/monthly-traffic-statistics/2012/august.aspx?sc_lang=en, accessed October 19, 2013.

EUROCONTROL (2013): Member States. http://www.eurocontrol.int/articles/member-states, accessed August 12, 2013.

European Commission (2011): Proposal for a Regulation of the European Parliament and of the Council on Common Rules for the Allocation of Slots at European Union Airports. EUR-Lex - Recherche Simple. http://eur-lex.europa.eu/LexUriServ/LexUriServ.do?uri=-CELEX:52011PC0827:EN:NOT, accessed August 10, 2013.

Germanwings (2006): Pressearchiv von Germanwings - Unternehmen. germanwings.com. http://www.germanwings.com/de/Unternehmen-Pressearchiv.htm, accessed October 21, 2013.

Germanwings (2013b): Boomerang Club – Sammeln Sie Meilen bei Ihrem Vielfliegerprogramm. germanwings.com. https://www.germanwings.com/skysales/Boomerang.aspx?culture=de-DE, accessed October 19, 2013.

Germanwings (2013c): Das Komfortbewegungsmittel. http://www.germanwings.com/de/-Service/BEST.htm, accessed October 16, 2013.

Lufthansa (2013b): Alliances and Partner Airlines. http://www.lufthansagroup.com/en/company/alliances.html, accessed October 6, 2013.

Miles & More (2013): Miles and More. http://www.miles-and-more.com/online/portal/mam/de/-earn/flight?l=de&cid=18002, accessed October 19, 2013.

n.a. (2009): Verband: Krise ändert Geschäftsreise-Praktiken. Airliners.de. http://-www.airliners.de/verband-krise-aendert-geschaeftsreise-praktiken/18409, accessed August 23, 2013.

n.a. (2013): Abschied von einem Klassiker: Die Tage der Boeing 747 sind gezählt - Luftfahrt. FOCUS Online. http://www.focus.de/finanzen/news/unternehmen/luftfahrt/abschied-von-einem-klassiker-die-tage-der-boeing-747-sind-gezaehlt_aid_986014.html, accessed August 28, 2013.

Ryanair (2013): Passenger Traffic 2002/2013. http://www.ryanair.com/de/investor/traffic-figures, accessed October 19, 2013.

Statista (2013): Prognose zur Entwicklung des realen BIP in Deutschland bis 2014. http://de.statista.com/statistik/daten/studie/73760/umfrage/entwicklung-des-realen-bip-in-deutschland-bis-2011/, accessed September 30, 2013.